Table of Contents

This book gives (about 10,000) different truss designs. There are 16,443 combinations of span, lumber grade, slope, etc, but loads below 12 psf and above about 100 psf are not included.

This book deals only with truss, roof and ceiling design and construction. The rest of the building—girders, lintels, walls, foundations, etc—must also be well designed and built for a building to perform as expected. Some publications that include related information are:

MWPS-1, MWPS Structures and Environment Handbook

A number of plans for specific buildings are also available. Write to one of the universities listed inside the front cover for a free catalog giving prices and descriptions.

This book gives the maximum roof load a given truss will carry in addition to the specified ceiling load.

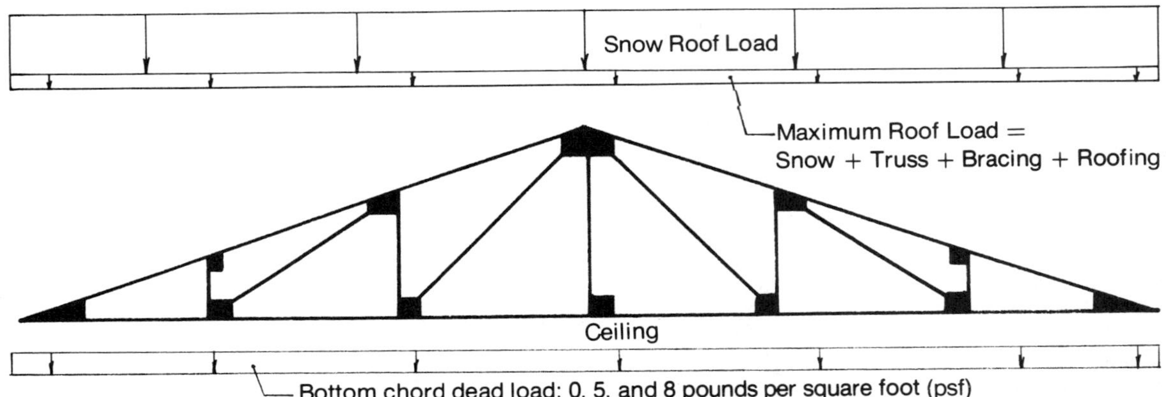

Parts of a Truss

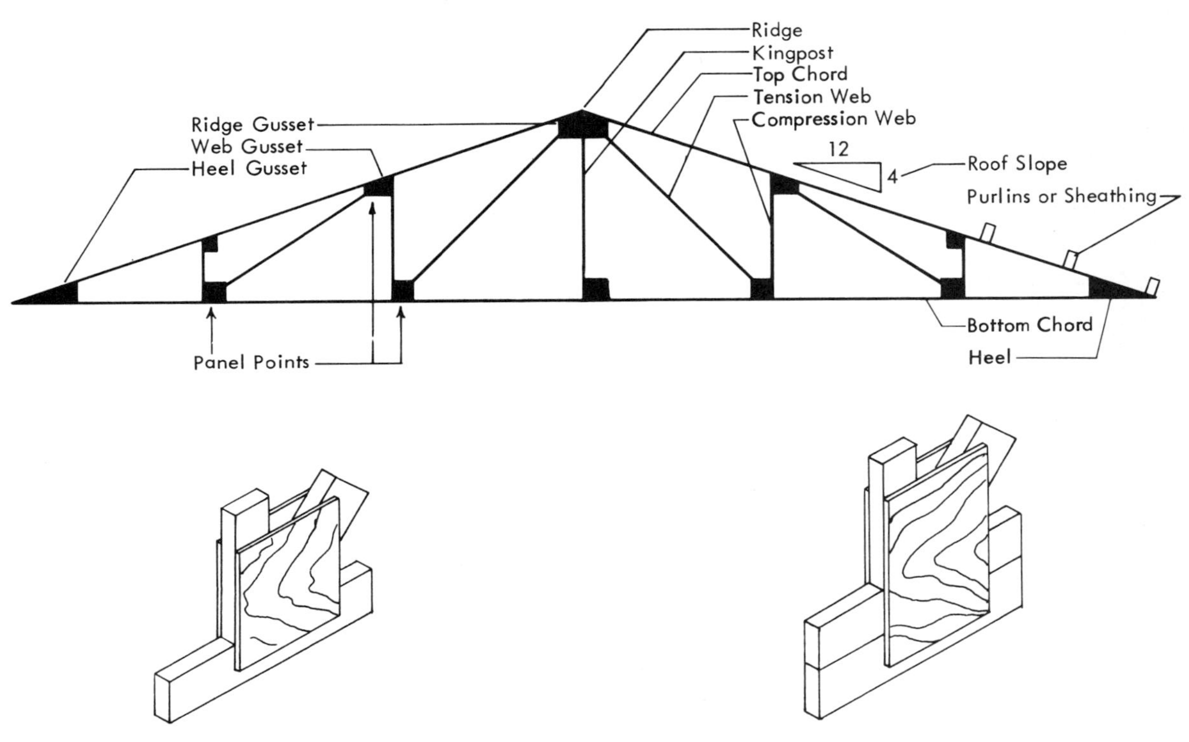

Gussets are on both sides of each joint

"Stacked" Bottom Chord

Some abbreviations used in this book:

ANSI—American National Standards Institute
MSR—Machine Stress Rated
NDS—National Design Specification for Wood Construction
PDS—Plywood Design Specification
PPSA—Purdue Plane Structures Analyzer
psf—pounds per square foot
psi—pounds per square inch
MWPS—Midwest Plan Service
NRAES—Northeast Regional Agricultural Engineering Service

This book discusses planning, building, and erecting clear-span gable roof trusses. Trusses can be used with any wall construction. Clear-span building construction provides for simple floor space arrangement. Halls, alleys, partitions, and equipment can be arranged, and rearranged, for best functional use of the space. There are no posts or bearing walls in the way. Machines and vehicles have maximum freedom of movement inside the building.

The bottom chords of trusses can support a ceiling. Head room clearance is restricted and bird roosting may be a problem compared to rigid frame construction.

READ and USE **all** parts of the book.

1. Make all decisions before construction starts. Record dimensions and other specifications on a **Summary Page** in the back of the book.
2. Select the right truss design: use the specifications on the following pages and the design tables for your span.
3. Build the trusses, using specifications and procedures under "Construction."
4. Erect the trusses as shown in the section on "Erection." Wind anchorage and bracing are essential parts of erection.

Truss Selection

Your truss selection depends on

- Web Pattern
- Span
- Roof Slope
- Truss Spacing
- Dead Loads
- Snow Load
- Wind Loads
- Lumber Type

Web Pattern

These truss designs use the Pratt web pattern, which is good for buildings with or without ceilings. The number of webs increases with heavier loads and longer spans for maximum use of chord lumber.

Up to 34' span, use 2-web trusses. Both 2- and 4-web designs are given for 36'-40' spans, and 4- and 6-web designs for 46'-50' spans. If your truss is in these ranges review both web patterns. Above 50' use 6-web trusses.

Increasing the number of webs increases the number of pieces of lumber and plywood, but may decrease chord sizes.

The three patterns are illustrated in Fig 1.

Span

Span is measured as in Fig 2.

For spans between those dimensioned, use member sizes, joints, etc., for the next larger span. Example: For a 29' span, use a 30' truss design, with shorter lumber lengths.

Limit overhangs to 2' because they cause increased bending in the top chord at the heel joint. For a 2' to 4' overhang, use the top chord and heel gusset design for 1/3 larger snow load. Fig 3a.

Trusses can extend beyond exterior walls, but not farther than the length of the bottom chord scarf cut. Fig 3b.

Roof Slope

Roof slope significantly affects the forces in truss members. For a given span, spacing, and load, the member forces increase as the roof slope is decreased.

- 3/12 slope is common in homes, commercial buildings, and some farm buildings in low snow load areas; also for short spans and close spacings.
- 4/12 is the most common in farm buildings.
- 5/12 slope is common for buildings in areas of high snow loads or for long spans and wide spacings, because the steeper slopes reduce snow loads and lower member stresses.

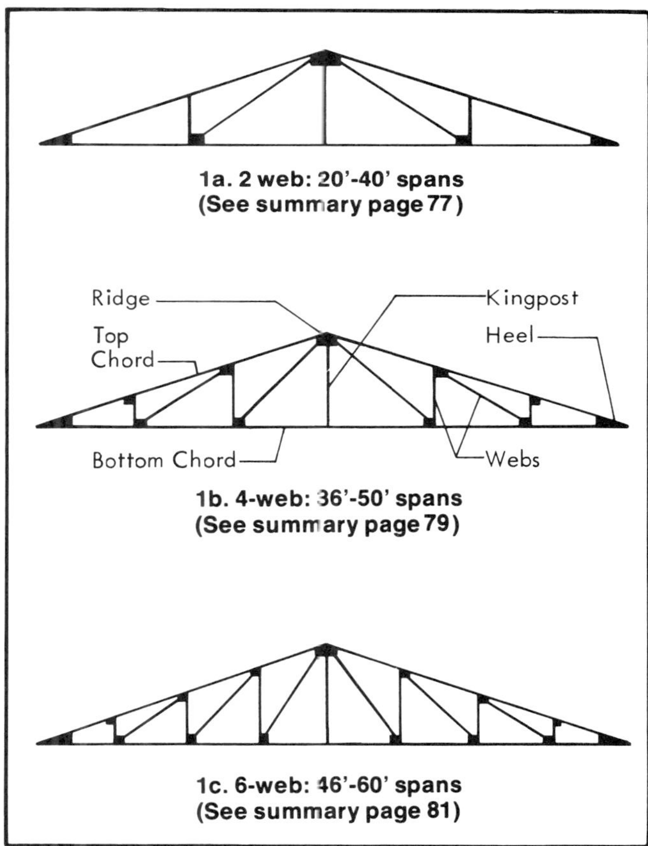

Fig 1. Web patterns.

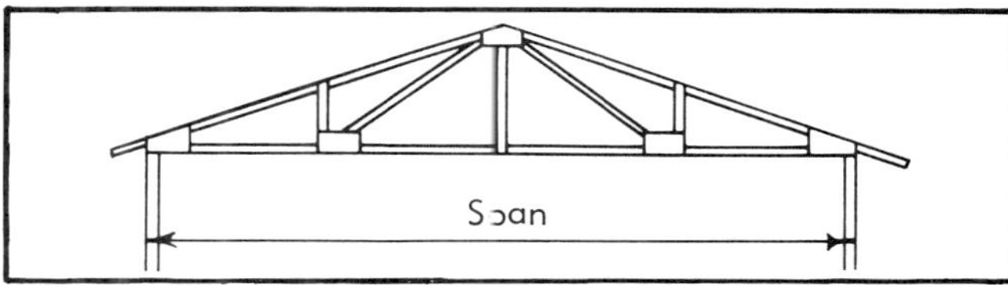

Fig 2. Clear span measurement.

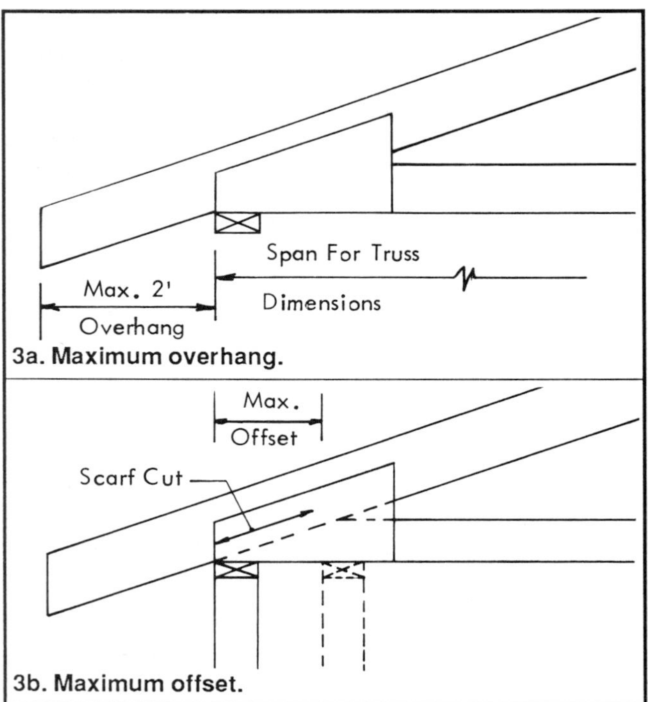

Fig 3. Overhang and offset.

Truss Spacing

Roof and ceiling materials, and wall framing, influence truss spacing selection. See "Roof and Ceiling Construction".

- 2' spacing uses more truss material and labor than wider spacings. It is common in short spans and low roof slopes for buildings with ceilings, solid roof decks, and shingles.
- 4' spacing supports many types of ceilings without special framing. It is common in insulated livestock buildings with metal roofs and in some machinery and equipment storage buildings.
- 8' spacing requires the least truss material and labor. It is common in buildings without ceilings: machinery and equipment storage, uninsulated livestock buildings, etc. Total cost may be greater if a ceiling is needed.

If no design is shown for the spacing you want at the load you must support, you may place 8'-design trusses 7' o.c., for example.

In a pole building, pole spacing can also affect truss spacing. With trusses 8'o.c., support each truss with a pole if possible. Specially designed girders can support trusses between poles.

Ceiling Dead Load

Three ceiling dead load cases are included:

- 0 psf allows for no materials in addition to the truss, bracing, and stiffeners.
- 5 psf ceiling dead load allows for a metal or plywood ceiling with insulation (warm livestock buildings).
- 8 psf ceiling dead load allows for a gypsum board ceiling with insulation (residential or light commercial buildings).

Add the weights of framing, insulation, and any fixtures supported by the ceiling to get the dead load on the bottom chord. Table 1.

Ceiling load note

Some trusses allow a load with 5 psf or even 8 psf ceiling load, but allow none for 0 psf ceiling load. This apparent error occurs under wind loading, when lower chords try to buckle sideways.

- Truss designs assume that with no ceiling load, truss lower chords are laterally braced only at the lower chord panel points. See the lower chord stiffeners, Fig 14, page 12.
- Truss designs for 5 psf or 8 psf ceiling loads assume that framing to support the ceiling also stiffens the lower chord against lateral buckling, which makes the truss stronger under wind load.
- If lower chords are braced the full length of the building with an additional brace midway between each pair of panel points, you may use the load allowed for 5 psf ceiling load for the 0 ceiling load.

Roof Dead Load

You will need to know roof dead load to select a truss from the design tables. Add the weights of the truss, purlins or decking, roofing, and roof insulation to get the dead load on the top chord. Tables 1 and 2.

Example: Insulated livestock building

Bottom chord dead load	
½" plywood ceiling	1.4 psf
6" of insulation	2.4 psf
	Total 3.8 psf
Top chord dead load	
Truss weight—Ex. Table 2	2.0 psf
2x4 purlins—2' o.c.	0.7 psf
28 ga steel roofing	0.9 psf
	Total 3.6 psf

Table 1. Weights of roofing and ceiling materials.

Roof framing	
2x4 purlins, 2' o.c.	0.7 psf
2x6 purlins, 2' o.c.	1.1
Ceiling framing	
1x3 furring, 16" o.c.	0.4 psf
2x4 furring, 2' o.c.	0.7
Sheathing, etc.	
1" lumber, solid	2.2 psf
⅜" plywood	1.1
½" plywood	1.4
0.024" aluminum	0.4
28 ga steel	0.9
Asphalt shingles	2.6
Insulation, per inch of thickness	0.1-0.4

Table 2. Approximate weights of trusses, psf.
Example: A 4-web truss for 4' spacing with 2x8 top chord and 2x6 bottom chord weighs about 1.3 + 0.7 = 2.0 psf. Underlines in table indicate example.

Chord size Top	Bottom	Truss spacing 2'	4'	8'
		Truss dead weight, psf		
2x4	2x4	1.6	0.8	0.4
2x6	2x4	2.0	1.0	0.5
2x6	2x6	2.4	1.2	0.6
2x8	2x6	2.7	1.3	0.7
2x10	2x4+2x4	3.3	1.6	0.8
2x12	2x4+2x6	4.0	2.0	1.0
2x12	2x6+2x6	4.4	2.2	1.1
Add the following for:				
2-&4-Web Truss		1.4	0.7	0.4
6 Web Truss		2.1	1.2	0.6

Snow Load

This book helps you find a truss strong enough for the loads in your area.

- If a building code applies, it is the law and must be followed. A building code specifies required design loads in most cities, many towns, and some counties and states.
- If no building code applies, consider local practice and a building code for a nearby city. Note that building codes may exclude family residences and farm buildings.
- The following material on snow loads is a guide.

The snow loads in Table 3 are estimated from the map, which shows snow on the ground. About 70% of the ground snow is expected on a roof. Farm buildings are often designed for lower risk, so a 25-yr rather than 50-yr recurrence interval is used. Drifting of snow from adjacent higher structures, trees, or roofs is not included.

Add the snow load to your roof dead load to get the total required roof load in pounds per square foot, psf. Look for a truss to carry at least the total roof load for the lumber quality you will use, and the span, slope, spacing, and ceiling dead load for your building.

Table 3. Recommended snow loads.
Farm buildings: 50-yr map load x 0.8 for 25-yr x 0.7 for snow on roof.
Other buildings: 50-yr map load x 0.7 for snow on roof.
Minimum recommended load is 12 psf + roof dead load.
In areas where all of the maximum snow load results from a single storm without significant wind, the maximum roof load may equal the ground snow load.

Map load	Roof snow load Farm	Other	Map load	Roof snow load Farm	Other
	psf			psf	
15	12.0	12	60	33.6	42
20	12.0	14	70	39.2	49
25	14.0	20	80	44.8	56
30	16.8	21	90	50.4	63
35	19.6	24.5	100	56.0	70
40	22.4	28	110	61.6	77
50	28.0	35	120	67.2	84

Wind Loads

All trusses in this book are designed to withstand winds of 80 miles per hour against a building in suburban areas, towns, city outskirts, wooded areas, and rolling terrain. Winds tend to lift the roof and roll the building over. See "Windbracing" and "Wind Anchorage". In open flat coastal belts, and in areas having winds higher than 80 miles per hour, provide additional anchorage and additional bracing for the bottom chord and long diagonal webs.

Truss anchorage and the design of each member allow for the worst condition--open or closed building and wind from any direction.

Concentrated Loads

These trusses have not been designed for large concentrated loads. Avoid cranes and heavy roof-mounted equipment.

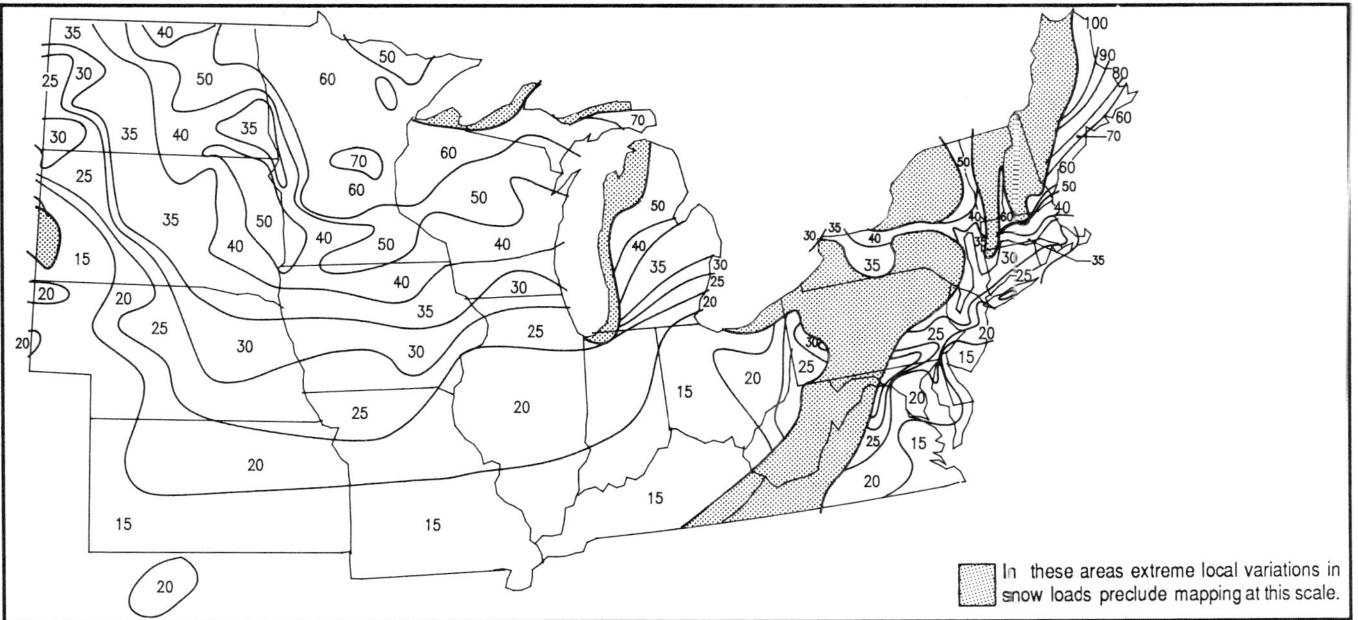

Fig 4. Snow loads on the ground, psf, 50-yr recurrence.
50-yr recurrence interval. Adapted from Standard EP288.4, American Society of Agricultural Engineers.

Materials

Lumber

Wood for trusses must be of uniform good quality and dried to at least 19% (In Fig 5, "S-dry" means 19%, "MC-15" or "KD" means 15%.). The designs are based on standard dry milled lumber sizes:

Top Chords

Truss designs use only 2x4s and 2x6s in the bottom chords because recent lumber grading rules have substantially reduced strength in tension of 2x8s, 2x10s and 2x12s. Do not substitute a 2x8, 2x10, or 2x12 for the specified stacked members. See MSR Lumber.

2x4 1½x3½ | 2x6 1½x5½ | 4+4 1½x7 | 4+6 1½x9 | 6+6 1½x11

Bottom Chords

Trusses are designed and tabulated for the three stress levels given in Table 4. Table 5 gives allowable design stresses for different species, grades, and sizes of lumber from NDS.

Table 4. Lumber stresses used in design.

Allowable design stresses (psi) are increased above those tabulated by 15% for repetitive bending, 15% for snow + dead loads, and 33% for wind + dead loads. Web and chord lumber used must meet these specifications. See Table 5.
Fb = bending. Ft = tension; Fc = compression parallel to grain;
E = modulus of elasticity.

Group, Fb	Ft	Fc	E
	psi		
1600f	1000	1250	1,700,000
1400f	825	975	1,300,000
1100f	625	825	1,100,000

Examples of Truss Selection:

1. You wish to use No. 2 Douglas Fir (North) milled and used at 19% moisture content. This lumber is in the 1100 group (Table 5a) for all member sizes. Select a truss for your span, slope, spacing, and loading from the 1100 group in the design tables.
2. Your total roof load is 34 pounds. For a 36', 4-web truss, 4/12 slope, with a 4' spacing on page 40, you find that the 1400 group has an allowable upper chord load of 37 psf (2x6 top and bottom chords), assuming a 5 psf/lb ceiling load. Any of the lumber grades and species in the 1400 and the 1600 group may be used for this truss.

   Note that 2x4s may not be in the same group as larger members.
3. You have selected a truss in the design tables from the 1600 group, and you plan to use Southern Pine at 15% moisture content. Use No. 2 Dense (or better) for 2x4 chords and No. 1 lumber for 2x6 and larger chords. If your lumber falls in two different stress groups (2x4s are in 1400 group and 2x6+ are in 1100 group), select your truss from the lower stress group.

Machine Stress Rated Lumber

Machine stress rated lumber has been evaluated by mechanical stress-rating equipment. MSR lumber is different from visually graded lumber because each piece is nondestructively tested and is marked to indicate the modulus of elasticity. MSR lumber also meets certain visual requirements. Trusses are a common application for MSR lumber.

Table 5B gives design values for some MSR lumber grades. Use these MSR design values and Table 4 to find the appropriate lumber group for selecting trusses from the design tables. Note that MSR E values are more "controlled" than visually graded lumber. You can expect 1.5E-1650 MSR to have less than 5% of its E values below about 1.2 while visual 1.7E, if it averaged 1.7, is expected to have 5% below 1.0 or about 13% below 1.2. This is apparent justification for using, for example, the 1.5E-1650 MSR lumber in the 1600 group, even

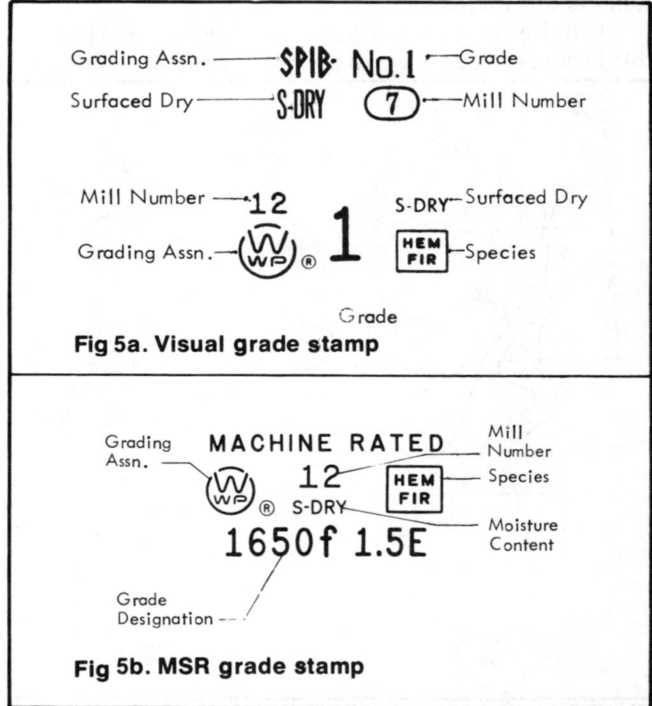

Fig 5a. Visual grade stamp

Fig 5b. MSR grade stamp

Fig 5. Lumber grade stamps

Table 5a. Allowable stresses for commercial lumber, psi.

Species, grades and sizes are grouped to aid truss selection. Source: Design Values for Wood Construction, 1986 Supplement to 1986 NDS.

2x6 + = 2x6, 2x8, 2x10, 2x12. (Ft applies only to 2x6)
SS = Select structural
(15%) = moisture content at time of milling

1600 Group

Species	Grade	Size	Fb	Ft	Fc	Ex 10^6
			psi			
Douglas Fir—Larch	No. 1	2x4	1750	1050	1250	1.8
	SS	2x6+	1800	1200	1400	1.8
Douglas Fir—Larch (North)	No. 1	2x4	1750	1050	1250	1.8
	SS	2x6+	1800	1200	1400	1.8
Southern Pine (15%)	No. 2 dense	2x4	1800	1050	1350	1.7
	No. 1	2x6+	1600	1050	1450	1.8
Southern Pine (19%)	No. 1	2x4	1700	1000	1250	1.7
	No. 1 dense	2x6+	1700	1150	1450	1.8

1400 Group

Species	Grade	Size	Fb	Ft	Fc	Ex 10^6
Balsam Fir	No. 1	2x4	1450	850	1050	1.5
Balsam Fir	SS	2x6+	1500	1000	1200	1.5
Coast Sitka Spruce	SS	2x4	1500	875	1100	1.7
Coast Species	SS	2x4	1500	875	1100	1.5
Douglas Fir—Larch	No. 2	2x4	1450	850	1000	1.7
	No. 1	2x6+	1500	1000	1250	1.8
Douglas Fir—Larch (North)	No. 2	2x4	1450	850	1000	1.7
	No. 1	2x6+	1500	1000	1250	1.8
Hem—Fir	No. 1	2x4	1400	825	1050	1.5
	SS	2x6+	1400	950	1150	1.5
Sitka Spruce	SS	2x4	1550	925	1150	1.5
Southern Pine (15%)	No. 2	2x4	1550	900	1150	1.6
Southern Pine (19%)	No. 2	2x4	1400	825	975	1.6
	No. 1	2x6+	1450	975	1250	1.7
Spruce—Pine—Fir	SS	2x4	1450	850	1100	1.5
Western Hemlock	No. 1	2x4	1550	900	1150	1.6
	SS	2x6+	1550	1050	1300	1.6

1100 Group

Species	Grade	Size	Fb	Ft	Fc	Ex 10^6
Aspen	SS	2x4	1300	775	850	1.1
California Redwood	No. 2	2x4	1400	800	1100	1.25
	No. 2	2x6+	1200	650	1200	1.25
Coast Sitka Spruce	No. 1	2x4	1250	750	875	1.7
	No. 1	2x6+	1100	725	875	1.7
Coast Species	No. 1	2x4	1250	750	875	1.5
	No. 1	2x6+	1100	725	875	1.5
Douglas Fir—Larch	No. 2	2x6+	1250	650	1050	1.7
Douglas Fir (North)	No. 2	2x4	1450	850	1000	1.7
	No. 2	2x6+	1250	650	1050	1.7
Douglas Fir (South)	No. 2	2x4	1400	825	900	1.3
	No. 2	2x6+	1200	625	950	1.3
E. Hemlock—Tamarack	No. 2	2x4	1250	725	850	1.1
	No. 1	2x6+	1300	875	1050	1.3
E. Hemlock—Tamarack (North)	No. 2	2x4	1250	725	850	1.1
	No. 1	2x6+	1300	875	1050	1.3
Eastern Spruce	No. 1	2x4	1200	700	825	1.5
Eastern White Pine	No. 1	2x4	1150	625	850	1.2
	SS	2x6+	1150	775	950	1.2

1100 Group, continued

Species	Grade	Size	Fb	Ft	Fc	Ex 10^6
			psi			
Eastern White Pine (North)	No. 1	2x4	1150	675	850	1.2
	SS	2x6+	1150	775	950	1.2
Eastern Woods	SS	2x4	1300	775	850	1.1
Engelmann Spruce—Alpine Fir	SS	2x4	1350	800	950	1.3
	SS	2x6+	1200	775	850	1.3
Hem—Fir	No. 2	2x4	1150	675	825	1.4
	No. 1	2x6+	1200	800	1050	1.5
Hem—Fir (North)	No. 1	2x4	1350	800	1050	1.5
Hem—Fir (North)	SS	2x6+	1350	900	1150	1.5
Hem—Fir (North)	No. 1	2x6+	1150	775	1050	1.5
Idaho White Pine	No. 1	2x4	1150	650	875	1.4
	SS	2x6+	1150	775	950	1.4
Lodgepole Pine	No. 1	2x4	1300	750	900	1.3
	No. 1	2x6+	1100	750	900	1.3
Mountain Hemlock	No. 1	2x4	1450	850	1000	1.3
	No. 1	2x6+	1250	850	1000	1.3
Mt. Hemlock—Hem-Fir	No. 1	2x4	1400	825	1000	1.3
	No. 1	2x6+	1200	800	1000	1.3
Northern Aspen	SS	2x4	1300	750	850	1.4
Northern Pine	No. 1	2x4	1400	825	975	1.4
	No. 1	2x6+	1200	800	975	1.4
Northern Species	No. 1	2x4	1150	675	825	1.1
	SS	2x6+	1150	750	900	1.1
Ponderosa Pine	No. 1	2x4	1200	700	850	1.2
	SS	2x6+	1200	825	950	1.2
Ponderosa Pine—Sugar Pine	No. 1	2x4	1200	700	850	1.2
	SS	2x6+	1200	825	950	1.2
Red Pine	No. 1	2x4	1200	700	825	1.3
	SS	2x6+	1200	775	900	1.3
Sitka Spruce	No. 1	2x4	1350	775	925	1.5
	No. 1	2x6+	1150	775	925	1.5
Southern Pine (15%)	No. 2	2x6+	1300	675	1200	1.6
Southern Pine (19%)	No. 2	2x6+	1200	625	1000	1.6
Spruce, Pine, Fir	No. 1	2x4	1200	725	875	1.5
	SS	2x6+	1250	825	975	1.5
Western Cedars	No. 1	2x4	1300	750	950	1.1
	No. 1	2x6+	1100	750	950	1.1
Western Cedars (North)	No. 1	2x4	1250	725	950	1.1
	SS	2x6+	1250	825	1050	1.1
Western Hemlock	No. 2	2x4	1300	750	900	1.4
	No. 1	2x6+	1350	900	1150	1.6
Western White Pine	No. 1	2x4	1150	675	875	1.4
	SS	2x6+	1150	750	975	1.4
White Woods (Western Woods)	SS	2x4	1350	775	950	1.1
	SS	2x6+	1150	775	850	1.1

Table 5b. Design values for machine stress rated (MSR) lumber, psi.

Normal loading conditions; 19% maximum moisture content; Lumber loaded on edge; 2" thick members or less, all widths. Source: Design Values for Wood Construction, 1986 Supplement to NDS.

Grade designation	Extreme fiber in bending "F_b" Single-member uses	Extreme fiber in bending "F_b" Repetitive-member uses	Tension parallel to grain "F_t"	Compression parallel to grain "F_c"	Modulus of elasticity "E"
	Design values in pounds per square inch				
900f-1.0E	900	1050	350	725	1,000,000
1200f-1.2E	1200	1400	600	950	1,200,000
1350f-1.3E	1350	1550	750	1075	1,300,000
1450f-1.3E	1450	1650	800	1150	1,300,000
1500f-1.3E	1500	1750	900	1200	1,300,000
1500f-1.4E	1500	1750	900	1200	1,400,000
1650f-1.4E	1650	1900	1020	1320	1,400,000
1650f-1.5E	1650	1900	1020	1320	1,500,000
1800f-1.6E	1800	2050	1175	1450	1,600,000
1950f-1.5E	1950	2250	1375	1550	1,500,000
1950f-1.7E	1950	2250	1375	1550	1,700,000
2100f-1.8E	2100	2400	1575	1700	1,800,000
2250f-1.6E	2250	2600	1750	1800	1,600,000
2250f-1.9E	2250	2600	1750	1800	1,900,000
2400f-1.7E	2400	2750	1925	1925	1,700,000
2400f-2.0E	2400	2750	1925	1925	2,000,000

though the E value for this grade designation is below that required for the 1600 group.

With MSR lumber, stacked bottom chords are not required; for example, use a 2x8 MSR piece instead of the 4+4 shown in the table. Allowable tension stresses do not drop with larger members as in visually graded lumber.

Plywood

The truss selection tables specify gussets of 3/8" or 1/2" plywood. Because 5/8" and 3/4" plywood do not have enough added shear strength to justify their cost for heel gussets and a long gusset is wasteful, two layers of 1/2" plywood are specified where needed. Glue one layer of a double gusset to the dimension lumber and then glue the second layer on top of it.

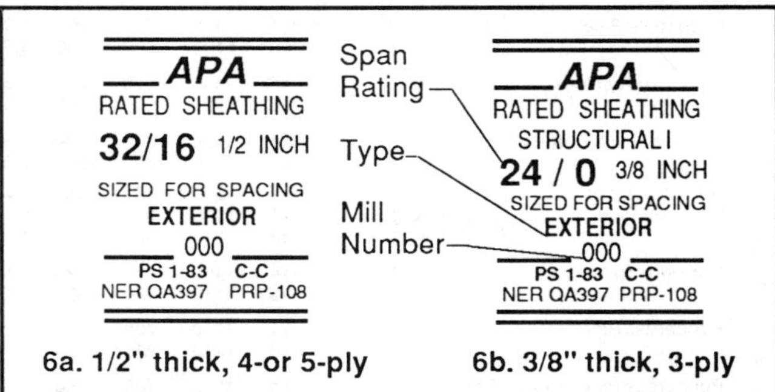

Fig 6. Plywood grade stamps.
See Table 7.

Plywood Requirements

Plywood quality is critical.

- C-C outer ply veneer grades. Or, CDX if building will remain dry, such as a home but not livestock housing--glue the C face.
- Exterior glue.
- 3/8" thickness: Use 3-ply plywood or COM-PLY, Structural I grade, unsanded APA Rated Sheathing.
- 1/2" thickness: Use 4-ply or 5-ply plywood, unsanded APA Rated Sheathing.
- Group 1 species in the outer plies.
- Do not use sanded or other sheathing grades.

Plywood stresses used in design: 250 psi shear through the heel joint. 53 psi rolling shear for glue area.

Glue

The right glue to use:

Casein that is highly mold and water resistant, though not water-proof, is adequate for trusses that will stay dry through their life. Resorcinol resin and epoxy resin glue are water-proof; use them for trusses with any joint exposed to unusual moisture conditions, such as in livestock buildings. Epoxy and casein glues are gap-fillers, that is they will fill in small spaces between lumber and plywood. Read the label carefully before you buy--make sure you're getting the right glue.

The right way to use it:

Follow the manufacturer's specifications for mixing, pot life (the length of time the glue stays good after mixing), temperature during use, etc.

Joints

The trusses in this book are designed for glued and nailed plywood gussets. Metal-plate trusses are becoming increasingly popular, but are designed somewhat differently, so direct substitution is not possible. Metal-plate trusses are usually designed by the plate or truss manufacturer. For further information, contact the Truss Plate Institute, 2400 East Devon Avenue, Des Plaines, Illinois 60018.

Gussets

The gussets dimensioned in the truss design tables fit over the members as shown in Figs 7 and 8. Apply a gusset to each side of each joint. All web and ridge gussets may be 3/8" plywood. Heel gussets are the length and thickness specified in the selection tables.

The heel gussets cover at least the scarf cut. **Face grain must be parallel to the bottom chord.**

Some pole builders fix the truss heel depth for sliding door header framing and offset the scarf cut. The heel gusset should cover the top chord a minimum of 4", (6" when two layers of 1/2" plywood are used).

Web gussets cover the top chord by 4" and the entire width of the vertical web. They cover 4" (3 1/2" for 2x4) of single bottom chords and 6" or 8" of stacked bottom chords. Do not move the gussets down to cover all of the bottom chord, or there will not be enough glue area on the webs.

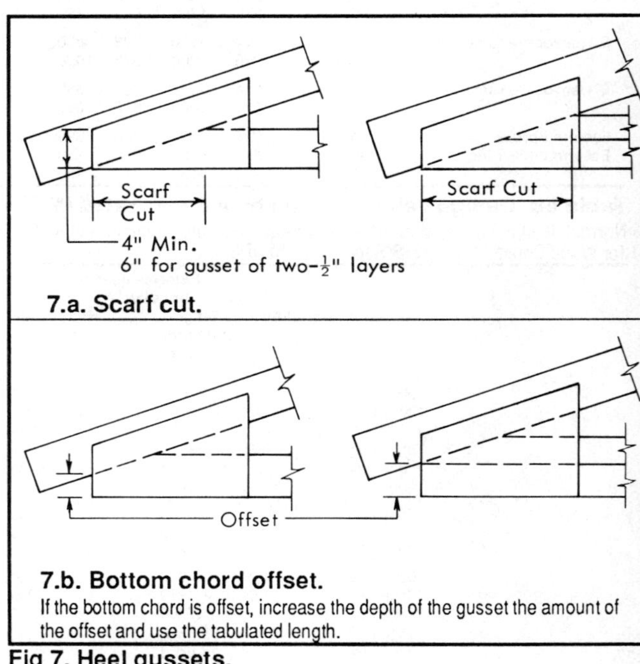

7.a. Scarf cut.

7.b. Bottom chord offset.
If the bottom chord is offset, increase the depth of the gusset the amount of the offset and use the tabulated length.

Fig 7. Heel gussets.

4"
H
W
Joint Ⓐ
Joint Ⓑ
Joint Ⓓ, Ⓕ
2x4 or
2x6
1"
2x6
2x6
Joints Ⓒ, Ⓔ, Ⓖ
Single Chord
Joints Ⓒ, Ⓔ, Ⓖ
Stacked Chord

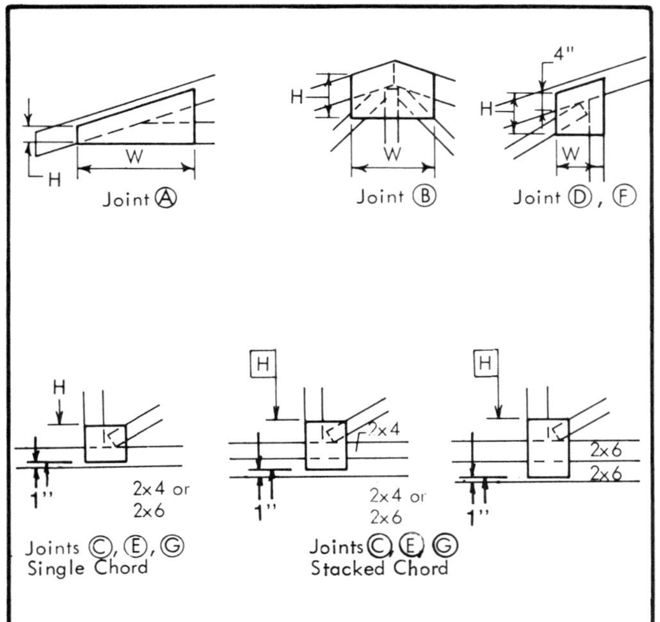

Fig 8. Tabulated gusset dimensions.

Ridge Gussets

Ridge gussets cover the ends of the webs and the full depth of the top chord.

Laps

Lap joints connect the shortest web to the top chord, and the king post to the bottom chord. Use 1" lumber or 3/8" plywood, as wide as the web or king post, glued and nailed to each side of each joint. Lap at least 4" over each member of the joint.

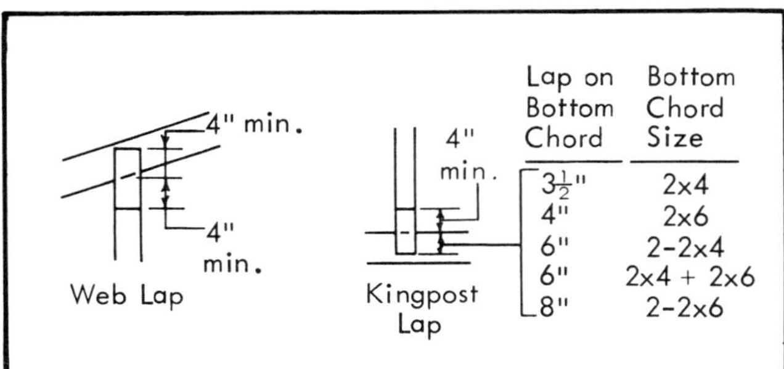

4" min.
4"
min.
Web Lap
4"
min.
Kingpost
Lap
Lap on Bottom Chord | Bottom Chord Size
3½" | 2x4
4" | 2x6
6" | 2-2x4
6" | 2x4 + 2x6
8" | 2-2x6

Fig 9. Laps.

Top and Bottom Chord Splices

Use full-length top chord and bottom chord lumber if possible.

When splices are necessary, locate them as shown in Fig 10. **Do not locate chord splices at the panel points because the combinations of forces from the chords and webs overload the joints.**

Build splices as shown in Fig 10.

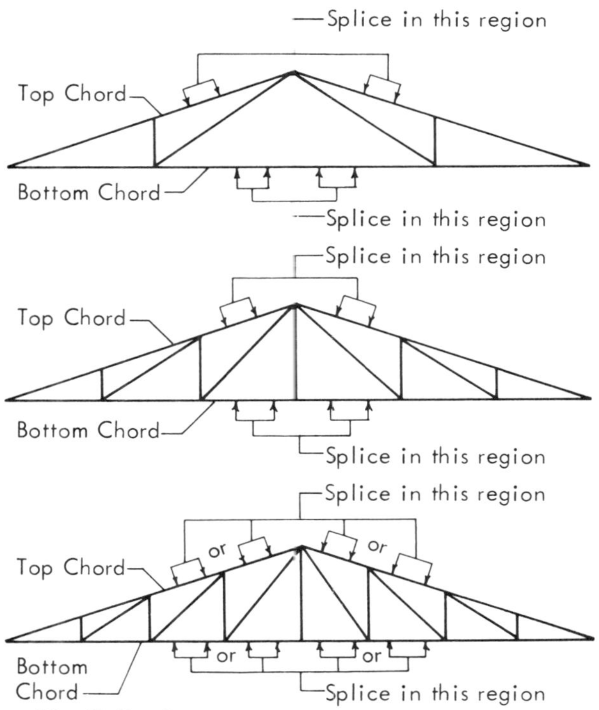

Splice in this region
Top Chord
Bottom Chord
Splice in this region
Splice in this region
Top Chord
Bottom Chord
Splice in this region
Splice in this region
Top Chord
or
or
Bottom Chord
or
or
Splice in this region

10a. Splice locations.

Glue and nail 1" lumber of same strength as chord on both sides
Chord
16" for 2x4
24" for 2x6 & up
Nails are 4" apart along grain, 2" apart across grain.
Glue and nail ½" plywood to both sides
2x4 Block
Chord
16" for 2x4
24" for 2x6 & up

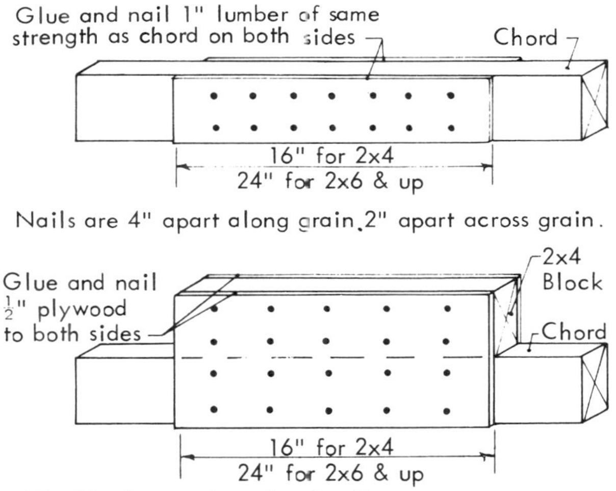

10b. Single member chord splice.
Face grain must be parallel to the chord.

End joint in chord
2x4x4 Block
Chord
12"
24'
12"
4'-0"

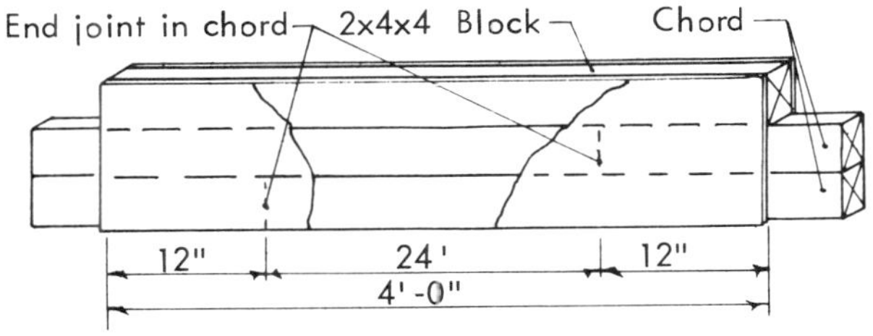

10c. Splices in stacked lower chords.
Offset end joints in lower chord members 24" min.
Cover one end joint in each chord member with one splice assembly.
(Part of plywood is cut away to show hidden members.)
Face grain must be parallel to the chord.

Fig 10. Chord splices.

Construction

Have the trusses made by an experienced truss builder if possible. If you build them yourself, learn how to make good glued joints.

Record dimensions and specifications on the appropriate Summary Page before starting construction.

Use clean, smooth lumber. Where members of different thicknesses (variation in manufacture) come together at a joint, sand or plane the thicker member so that the gusset can make full contact. Do not use cupped or twisted lumber. A good glue joint is essential for desired truss strength.

Use full-length top chord lumber if possible. Bottom chord splices are shown in trusses longer than 20'. If additional splices are necessary, place them as shown under "Splices."

Assemble One Truss

Select member sizes and lengths, and gussets required, from the design tables for your truss. **Note:** web members (W1, etc.) are given full length and will need to be cut to exact size.

A "camber" or slight raising of the bottom chord at the center of the truss provides for settlement from dead loads without the truss appearing to sag. Provide about 1/8" rise for each 10' of truss span.

Draw the complete truss on the floor or a plywood base. Cut lumber to fit. Lightly nail **(don't glue)** the complete truss together. Check the dimensions of the truss; both sides of the truss should be the same (rise, slope, position of joints).

Make a Jig

Nail 2x4 blocks to the base or floor around the lightly nailed truss. Mark off the 4" nail spacings along the jig at each joint.

Cut Other Trusses

Take the lightly nailed truss apart, and use its pieces as patterns for the rest of the trusses. Cut out the plywood gussets. The face grain of heel gusset "A" must be parallel to the bottom chord. The face grain of splices must be parallel to the chord. The direction of the grain in the other gussets is not critical.

Double Webs

Web members can buckle under snow loads or wind uplift in many large trusses—two 2x4s are specified to increase stiffness.

Build the truss with single web members, then cut an additional 2x4 to fit between the gussets. Nail it securely with 10d nails about 6" apart.

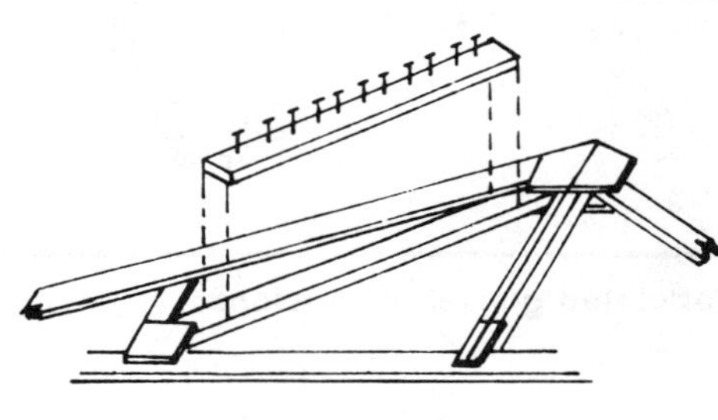

Double Web

In even larger trusses, both buckling due to wind uplift and tension due to snow loads are critical—a 2x6 and a 2x4 are specified. The 2x6 is required for tension and should be glued into the truss as the primary web member and the 2x4 added as a double web to increase stiffness. The 2x4s used as double webs are stiffeners and do not need to be of high quality.

Assemble and Glue

Use the jig to make uniform trusses.

Use plenty of glue—it's a low cost part of the building. Spread the glue with a fiber glue brush or paint roller.

Apply glue to both surfaces to be joined. When you

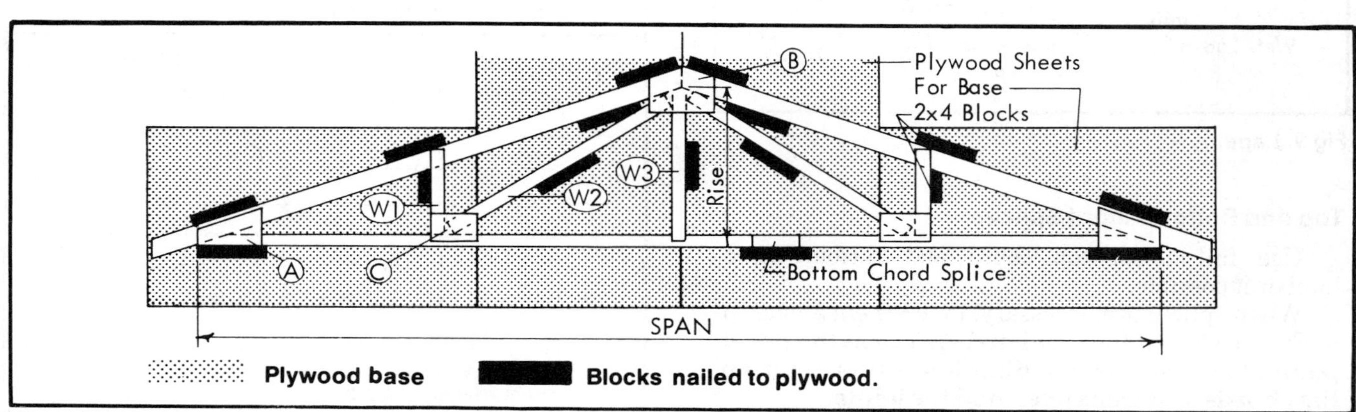

Fig 11. Truss jig.

nail the gussets in place, glue should ooze out from around the joint—if it doesn't, you're not putting on enough glue.

Use 5d or 6d box nails, preferably galvanized or cement coated. Space rows of nails 2" apart across the lumber grain. Nail 4" apart with the lumber grain, starting with a middle row, and nailing outward to the edges of the lumber. You may machine nail, but as each joint is nailed, hit each nail at least once with a hammer. Machines drive the fasteners so rapidly that the glue might not have time to spread—the final manual blow assures a tight joint.

Machine stapling is not recommended because failures have occurred.

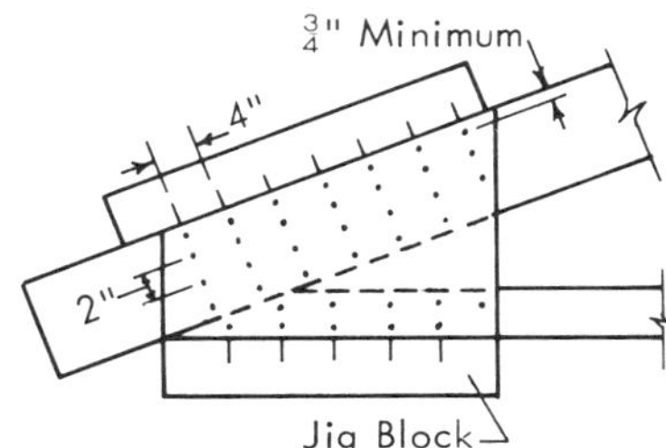

Nailing diagram

Raise the peak of the truss and turn the truss over, out of the jig.

Glue and nail the gussets on this side, and nail on the double webs if required.

To eliminate errors in the pattern, which may result in an uneven ridge, mark one end—e.g. right end—of each truss and erect the trusses with that end on the same side of the building.

Move the truss out of the way and store flat for at least 24 hours. After the glue has cured, the truss is ready to be placed on the structure.

Apply glue with a brush or roller to both surfaces.

Stack trusses flat to prevent warping, and in a dry place to cure.
Put spacers between the trusses to prevent sagging in the stack.

Precautions

Remove all dirt, oil, and sand from the lumber and plywood. This is necessary for a good glue bond.

Protect the glued joints from moisture for one week after fabrication. A plastic covering over stacked trusses is adequate.

Temperatures below 70 F delay curing. Trusses are ready to erect in 24 hours at 70 F, but require at least a week at 40 F.

Above 85 F, pot life may be very short. Keep the glue in the shade, and close and nail joints right after spreading. Consider keeping the pail of mixed glue in a bucket of ice water in hot weather.

Endwall Trusses

Endwall trusses are sometimes built without gussets on the outside face to simplify applying endwall siding. However, the endwall truss is not strong enough to support additional building length in any future expansion. If you might lengthen the building someday, put gusset plates on both sides of the joints and space plywood scraps along the truss members to apply siding.

Erection

Placing

Lift lighter trusses by hand and place them on the walls upside down. Lift heavier ones into place with a hoist. Keep the trusses as straight as possible during erection to reduce stress on the joints.

Position, anchor, and brace the trusses one at a time. Bracing helps align and hold trusses in place while the roof and ceiling are applied. Add anchorage, bracing, stiffeners, and purlins as you put the trusses up to prevent collapse during construction and windstorms.

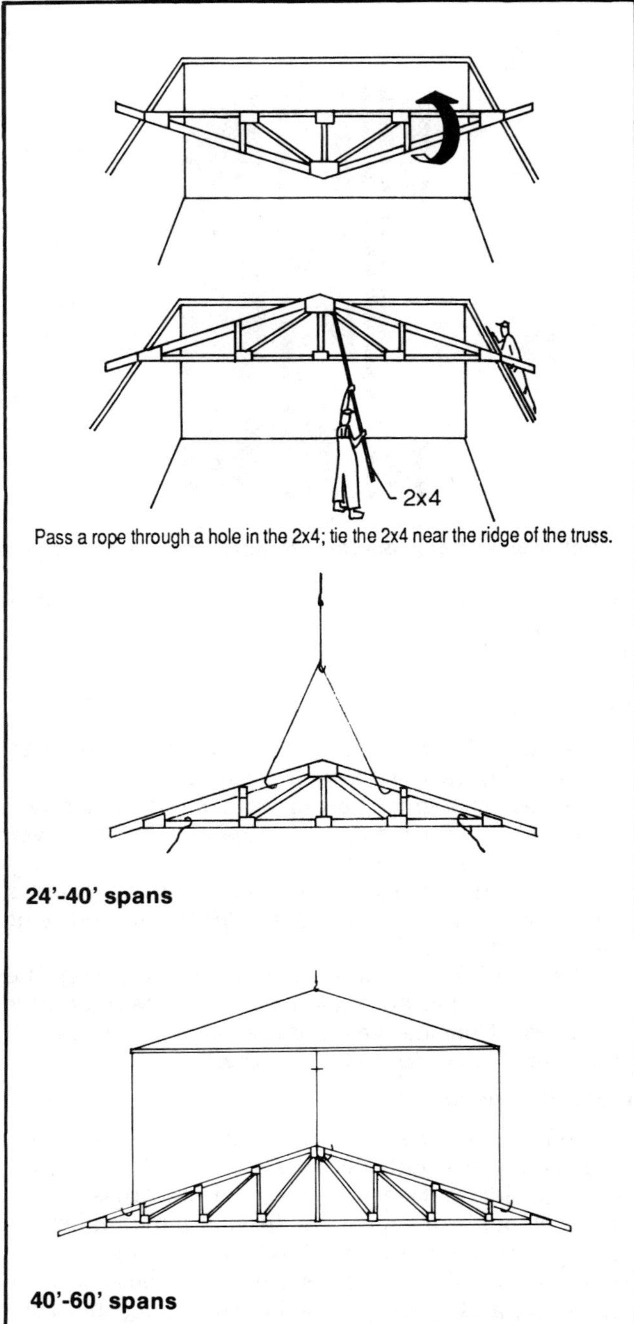

Fig 12. Truss placement

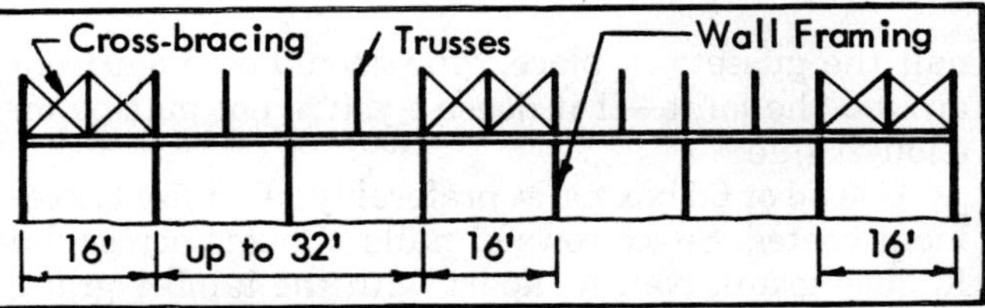

Fig 13. Cross-bracing.

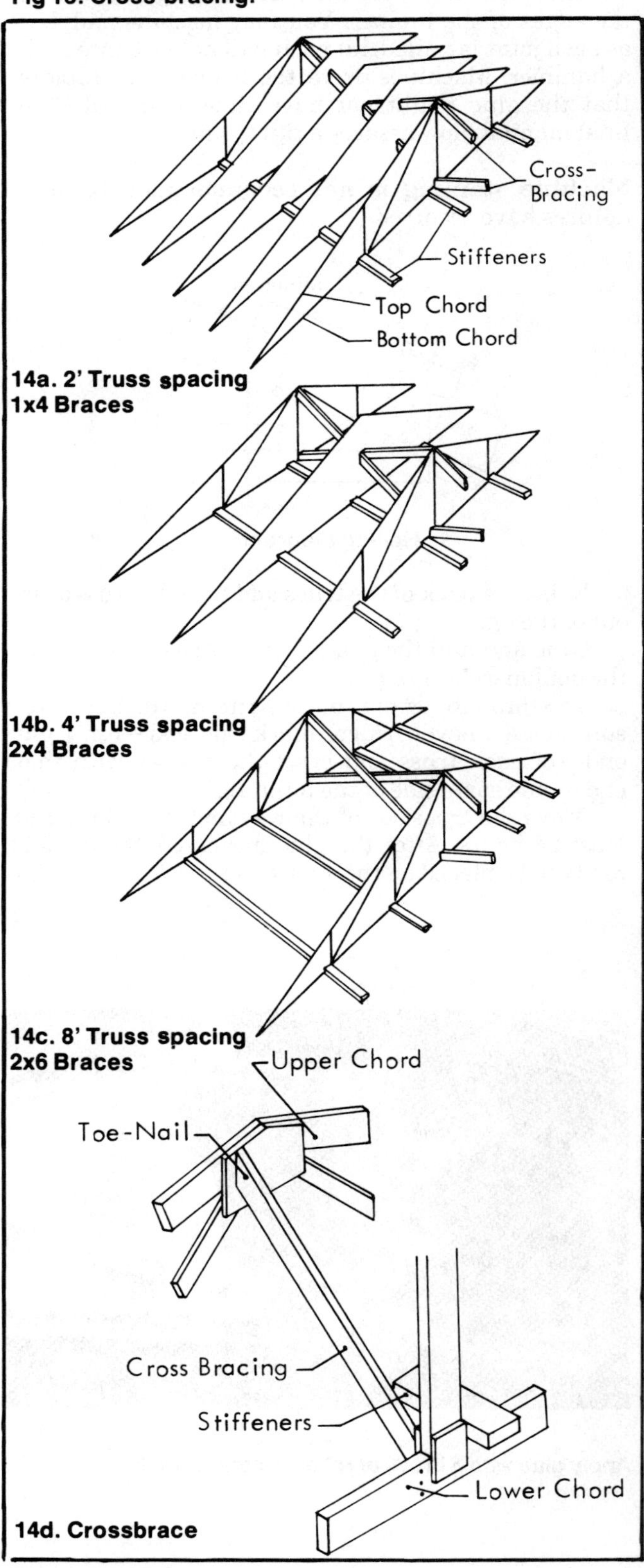

Fig 14. Windbracing.

Windbracing

Prevent wind damage during construction and for the life of the building with windbracing, anchorage, and knee bracing.

Install stiffeners along the bottom chords at all panel points (where webs meet bottom chord) continuously from endwall to endwall unless a rigid ceiling is to be installed. Cross bracing on the king posts is needed in all buildings.

Install cross bracing for at least 16' at each end of the building. Space additional 16' braced sections not farther than 32' apart.

Wind Anchorage

The wall holds the truss up. Anchorage holds the truss down during wind loads.

Table 6. Fasteners for wind anchorage.

	---Truss Spacing---		
Truss Span	2'	4'	8'
20'-24'	1A or 1B	1A or 1B	2A or 1B
26'-30'	1A or 1B	1A or 1B	2A or 2B
32'-46'	1A or 1B	2A or 1B	3A or 2B
48'-50'	1A or 1B	2A or 1B	4A or 2B
52'-60'	1A or 1B	2A or 2B	4A or 3B

Minimum for both ends of each truss.
A=Framing Anchor B=½" Bolt
4-30d ringshank nails equal one ½" bolt.

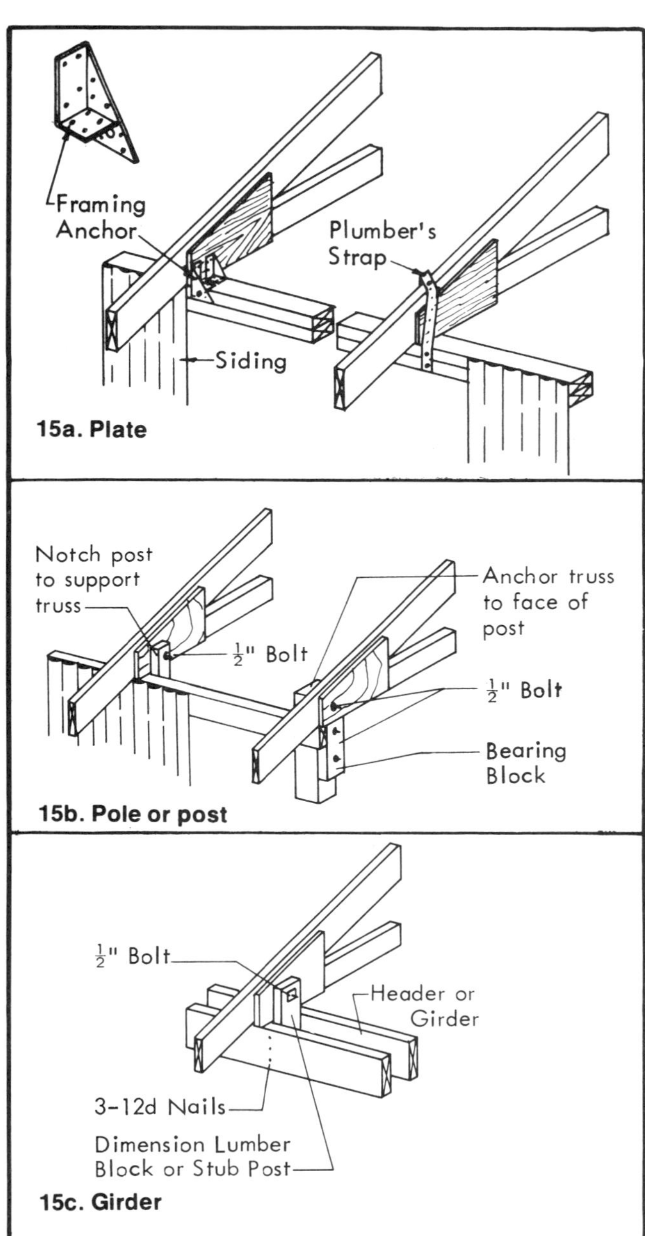

Fig 15. Truss anchorage.

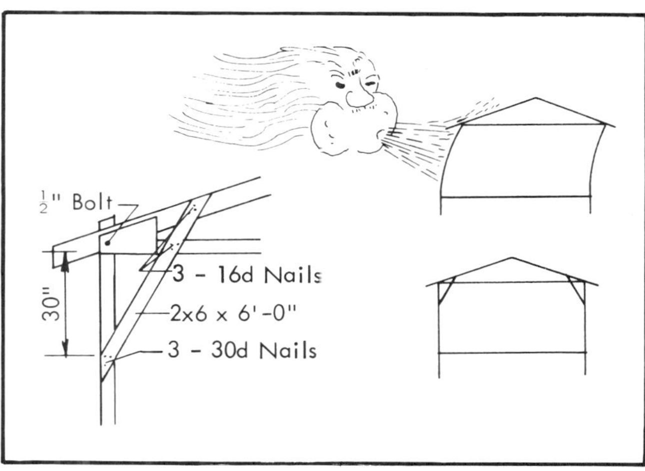

Fig 16. Typical knee brace—8' o.c.

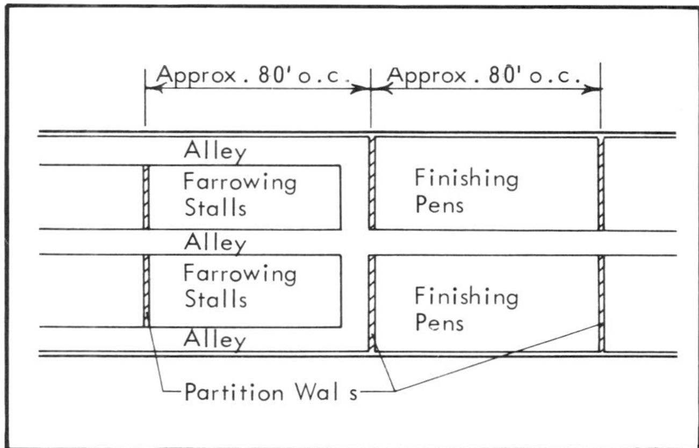

Fig 17. Windbracing with partition walls.

Knee Bracing

Knee braces stiffen the wall and roof joint to resist high winds and also help anchor the trusses. In buildings with rigid cross partitions, such as homes, knee bracing is not required. In pole buildings with poles sized assuming rigid pole-to-truss joints, the knee brace is part of the structural system. Knee bracing may be omitted in pole buildings with poles sized assuming pinned pole-to-truss joints.

In stud wall buildings where knee braces are not wanted, partition walls can be the wind bracing. Brace walls approximately every 80' by a partition wall of a structurally sound skin material: preferably plywood, possibly metal roofing or cement asbestos board; but not insulation or fiber board. Anchor the wall to the floor and to the bottom chord of a truss. Some small doorways will not seriously reduce the partition's stiffness.

Roof and Ceiling Construction

Roof Construction

After the trusses have been placed, anchored, and braced, apply the roof and ceiling.

Select the type of roofing and ceiling material to be used. Framing and sheathing required depends on the roofing material and the truss spacing.

Sheathing

Many types of material can be used as sheathing. Plywood is common. Table 7 and Fig 6.

The "Identification Index" is a design guide and is printed on the grade stamp on each sheet. The number to the left of the slash in the identification index is the maximum recommended support spacing for roof sheathing in residential construction. The number to the right is the spacing to use for subflooring.

Use 6d nails with sheathing 1/2" or less thick and 8d nails with thicker sheathing. Space nails 6" apart along panel edges; 12" apart along other supports. If panel supports are 48" o.c., space all nails 6" apart.

For support spacings of 32" or more, use T&G plywood, blocking under joints, or plyclips on panel edges between supports.

Plywood may expand with moisture; leave 1/16" space between roof or wall panel ends and 1/8" between sides to allow for possible expansion.

Support 1" lumber sheathing 24" o.c., 48" o.c. if it is T&G, in both farm and residential construction. Anchor with 8d nails. Space supports for rolled steel and aluminum roofing according to manufacturer's recommendations.

Table 7. Plywood roof sheathing for farm buildings.
Use plywood with the listed, or a larger, identification index. C-C, or CD, grade applied with face grain across supports.

	Support Spacing							
	16" o.c.		24" o.c.		32" o.c.		48" o.c.	
Snow Load	Ident. Index	Panel Thickness	Ident. Index	Panel Thickness	Ident. Index	Panel Thickness	Ident. Index	Panel Thickness
12	12/0	5/16"	16/0	5/16", 3/8"	24/0	3/8", 1/2"	32/16	1/2", 5/16"
18	12/0	5/16"	20/0	5/16", 3/8"	24/0	3/8", 1/2"	42/20	5/8", 3/4", 7/8"
24	12/0	5/16"	20/0	5/16", 3/8"	32/16	1/2", 5/8"	42/20	5/8", 3/4", 7/8"
Residential	12/0	5/16", 3/8"	24/0	3/8". 1/2"	32/16	1/2", 5/8"	48/24	3/4", 7/8"

Table 8. Maximum spans for roof purlins.
Maximum spans in feet for 1100f lumber. Top chords of trusses must be laterally supported 24" or less on center by the purlins or other framing.
Reduce purlin spacing if required by strength of roofing material.

Purlin Size	Purlin Spacing (in)	15	20	25	30	35	40	45	50	60	70	80	90	100
		Vertical Roof Load, Snow Plus Dead, Psf — Maximum Truss Spacing, Ft.												
1x4 Flat	16	3.6	3.3	3.1	2.8	2.6	2.4	2.3	2.2	2.0				
	20	3.4	3.1	2.8	2.5	2.3	2.2	2.1	2.0	1.8				
	24	3.2	2.8	2.5	2.3	2.1	2.0	1.9	1.8	1.6				
	28	2.8	2.4	2.2	2.0	1.8	1.7	1.6	1.5	1.4				
	32	2.6	2.3	2.0	1.9	1.7	1.6	1.5	1.4	1.3				
2x4 Flat	16	7.3	6.6	6.1	5.6	5.2	4.9	4.6	4.4	4.0				
	20	6.8	6.1	5.5	5.0	4.7	4.4	4.1	3.9	3.6				
	24	6.4	5.6	5.0	4.6	4.3	4.0	3.8	3.6	3.3				
	28	5.6	4.9	4.4	4.0	3.7	3.4	3.2	3.1	2.8				
	32	5.3	4.6	4.1	3.7	3.4	3.2	3.0	2.9	2.6				
2x4 Edge	16				8.6	8.0	7.5	7.0	6.7	6.1	5.6	5.3	5.0	4.7
	20			8.4	7.7	7.1	6.7	6.3	6.0	5.4	5.0	4.7	4.4	4.2
	24		8.6	7.7	7.0	6.5	6.1	5.7	5.4	5.0	4.6	4.3	4.1	3.9
	28		7.4	6.7	6.1	5.6	5.3	5.0	4.7	4.3	4.0	3.7	3.5	3.3
	32	8.0	7.0	6.2	5.7	5.3	4.9	4.6	4.4	4.0	3.7	3.5	3.3	3.1
2x6 Edge	16			14.8	13.5	12.5	11.7	11.1	10.5	9.6	8.9	8.3	7.8	7.4
	20			13.3	12.1	11.2	10.5	9.9	9.4	8.6	7.9	7.4	7.0	6.6
	24			12.1	11.1	10.2	9.6	9.0	8.6	7.8	7.2	6.8	6.4	6.1
	28			10.5	9.5	8.8	8.3	7.8	7.4	6.7	6.2	5.8	5.5	5.2
	32			9.8	8.9	8.3	7.7	7.3	6.9	6.3	5.8	5.5	5.2	4.9
2x8 Edge	16													
	20													8.7
	24											8.9	8.4	8.0
	28										8.2	7.7	7.3	6.9
	32									8.3	7.7	7.2	6.8	6.4

Roof Purlins

The maximum span for roof purlins depends on the sheathing or roofing materials used, and on the strength of the purlin itself. The maximum span between trusses, based on roof load and purlin spacing, is in Table 8.

Stagger purlin joints for continuity across the trusses. Purlins may be laid flat with 2' and 4' truss spacings. Butt ends over a truss and use 8', 12', or 16' lengths.

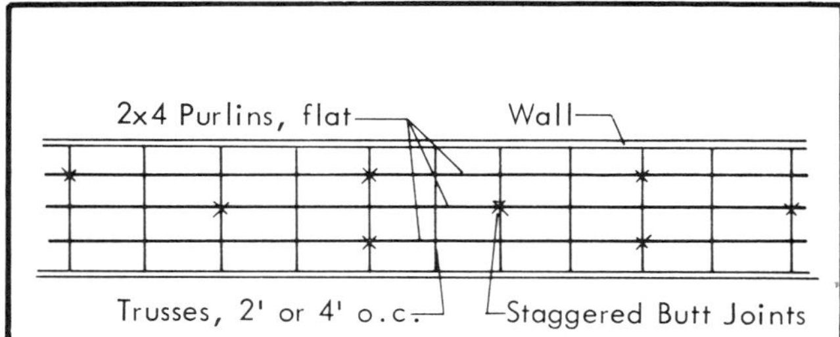

Fig 18. Staggered purlin joints for trusses 2' or 4' o.c.

For trusses 8' o.c., use 9', 16', and 18' purlins. Alternate 9' and 18' purlins up the roof at one endwall. Then alternate 16' and 18' purlins along the building, lapping the ends. Fasten the purlins to the trusses, and put four 10d nails through each lap.

Fig 19 illustrates mounting the trusses on poles which are 8' o.c.; note that the trusses are not all on the same sides of the poles so that 16' purlins fit better. Before construction starts, lay out exact truss locations on a floor plan of the building that shows large doors or other features that affect truss, purlin, and knee brace locations. The 16' purlins extend to the centerlines of the poles.

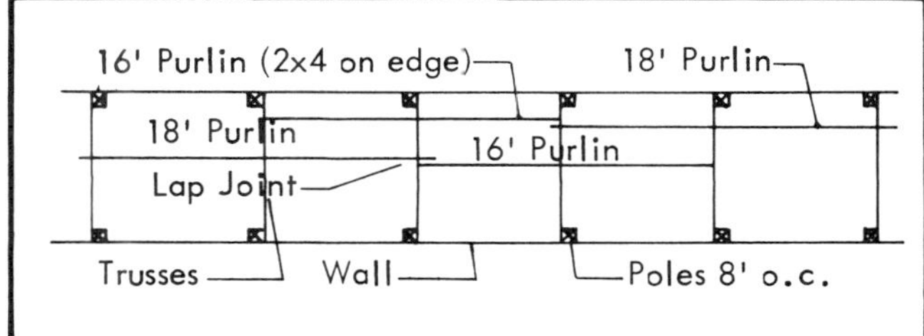

Fig 19. Staggered purlin joints for trusses 8' o.c.

Ceilings

Examples are given in Fig 21 for plywood ceilings. Rolled steel or aluminum roofing with the ribs running across the supports (as shown for plywood face grain) is about equivalent to ½" plywood and can support insulation above. Follow manufacturer's recommendations.

Leave spaces, 1/16" on ends and ⅛" on sides between plywood sheets in ceilings of high-humidity buildings, such as livestock housing, to prevent possible buckling from moisture expansion. Paint (exterior oil base) the edges of plywood sheets before installing them to prevent moisture movement in through the edges.

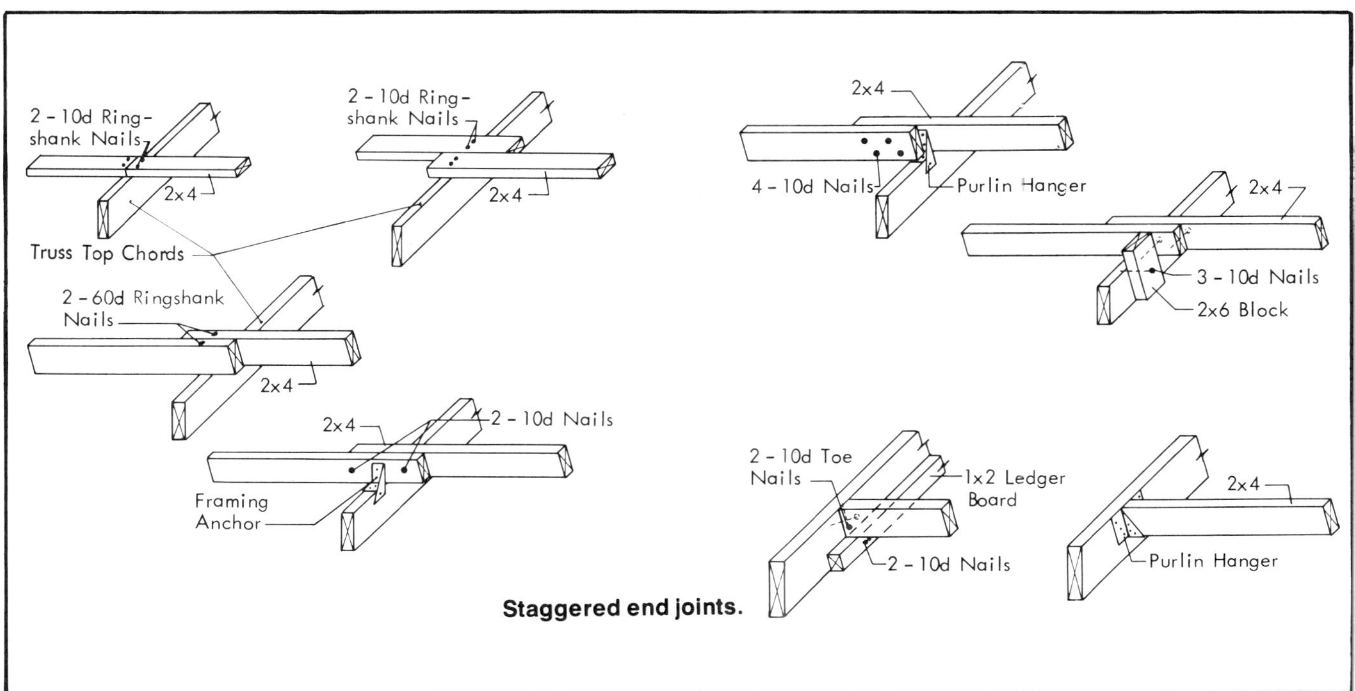

Fig 20. Fastening roof purlins.

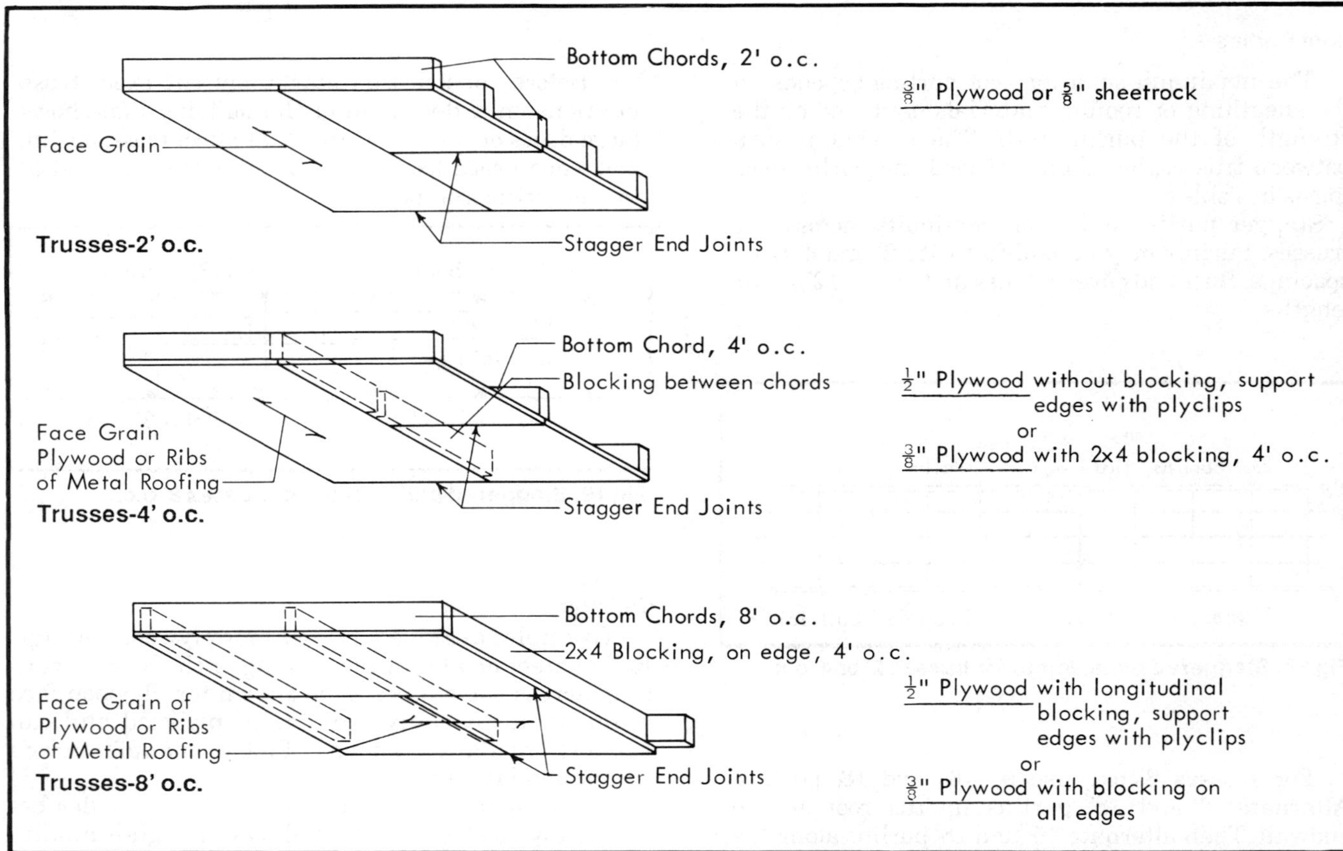

Fig 21. Ceiling construction.

NOTES

20' SPAN, 2-WEB

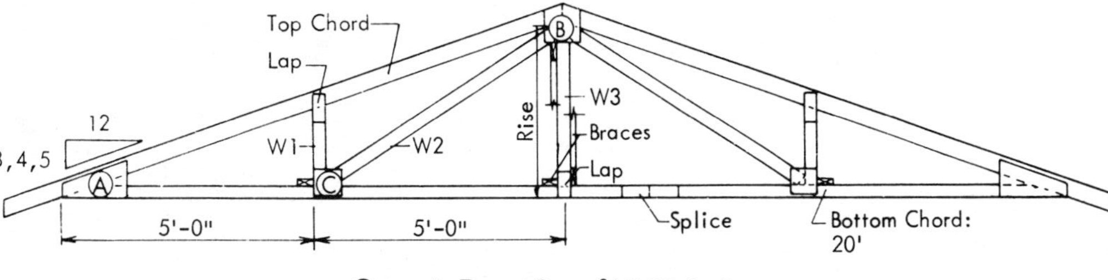

Gussets B and C are 3/8" thick plywood.

4+4, 4+6, 6+6 indicates stacked lower chord.
4&4, 6&4, indicate double web; a 2x4 is attached to the web member to increase its stiffness.

Before selecting heel gusset A, see pages 8 and 9 .

Web Lengths

Roof Slope	Rise	Top Chord	W1	W2	W3
3/12	2'-6"	11'	1'	6'	3'
4/12	3'-4"	11'	2'	6'	3'
5/12	4'-2"	12'	2'	6'	4'

1100f Lumber

			Truss spacing, ft. 2'			4'			8'			Web member sizes			Gusset Sizes, in. A	B	C
			Ceiling dead load, psf														
	Top chord	Bottom chord	0	5	8	0	5	8	0	5	8	W1	W2	W3	T H W	H W	H W
			---Max. snow + roof dead load, psf---														
3/12 Slope	2x4	2x4	41	39	36	18	15	0	0	0	0	2x4	2x4	2x4	3/8x3½x16	8x12	8x8
	2x6	2x4	68	66	62	29	25	0	14	0	0	"	"	"	3/8x4x23	10x12	"
	2x6	2x6	76	71	68	33	29	27	16	12	0	"	"	"	3/8x4x26	"	"
	2x8	2x6	95	86	87	41	36	33	20	15	0	2x4	2x4	2x4	3/8x4x31	12x16	8x10
	2x10	4+4	100+	100+	100+	66	58	54	33	25	0	"	"	"	½x4x28	14x16	10x10
	2x12	4+6	-	-	-	83	75	71	41	36	30	"	"	"	3/8x4x48	16x16	12x10
	2x12	6+6	-	-	-	82	73	69	41	35	32	"	"	"	3/8x4x49	"	14x10
4/12 Slope	2x4	2x4	47	45	43	20	18	0	0	0	0	2x4	2x4	2x4	3/8x3½x14	8x12	8x8
	2x6	2x4	90	86	85	39	33	0	19	0	0	"	"	"	3/8x4x23	10x12	8x10
	2x6	2x6	89	83	80	39	35	33	19	16	13	"	"	"	3/8x4x24	12x12	"
	2x8	2x6	100+	100+	100+	58	54	52	29	25	17	2x4	2x4	2x4	½x4x18	14x12	8x10
	2x10	4+4	-	-	-	82	74	70	41	33	0	"	"	"	½x4x22	16x12	12x10
	2x12	4+6	-	-	-	100+	99	95	52	47	42	"	"	"	½x4x28	16x16	14x10
	2x12	6+6	-	-	-	-	-	94	51	47	44	"	"	"	½x4x34	"	16x10
5/12 Slope	2x4	2x4	52	49	48	22	20	12	0	0	0	2x4	2x4	2x4	3/8x3½x13	8x12	8x8
	2x6	2x4	100+	96	96	44	42	13	22	0	0	"	"	"	3/8x4x21	10x12	"
	2x6	2x6	-	93	89	43	40	38	21	18	16	"	"	"	3/8x4x21	"	"
	2x8	2x6	-	100+	100+	64	60	57	32	28	21	2x4	2x4	2x4	½x4x17	12x16	8x10
	2x10	4+4	-	-	-	89	82	81	44	39	15	"	"	"	½x4x20	14x16	10x10
	2x12	4+6	-	-	-	100+	100+	100+	58	53	49	"	"	"	½x4x25	16x16	12x10
	2x12	6+6	-	-	-	-	-	-	56	51	48	"	"	"	½x4x28	"	16x10

1400f Lumber

Slope	Top chord	Bottom chord	2' — 0	2' — 5	2' — 8	4' — 0	4' — 5	4' — 8	8' — 0	8' — 5	8' — 8	W1	W2	W3	A T H W	B H W	C H W
			Truss spacing, ft. — Ceiling dead load, psf — Max. snow + roof dead load, psf									Web member sizes			Gusset Sizes, in.		
3/12 Slope	2x4	2x4	51	48	46	22	19	17	0	0	0	2x4	2x4	2x4	3/8x3½x20	8x12	8x8
	2x6	2x4	89	87	87	39	37	20	19	0	0	"	"	"	½x4x17	10x12	"
	2x6	2x6	94	87	83	41	37	34	20	16	14	"	"	"	3/8x4x32	10x16	8x10
	2x8	2x6	100+	100+	100+	54	49	46	27	22	19	2x4	2x4	2x4	½x4x23	12x16	8x10
	2x10	4+4	–	–	–	87	77	75	43	37	22	"	"	"	½x4x29	14x16	10x10
	2x12	4+6	–	–	–	100+	100	96	53	47	43	"	"	"	½x4x36	16x16	12x10
	2x12	6+6	–	–	–	–	90	90	50	45	40	"	"	"	½x4x44	"	14x10
4/12 Slope	2x4	2x4	58	55	54	25	23	21	12	0	0	2x4	2x4	2x4	3/8x3½x17	8x12	8x8
	2x6	2x4	100+	100+	100+	50	46	30	25	0	0	"	"	"	½x4x16	12x12	8x10
	2x6	2x6	–	–	–	48	44	42	24	20	18	"	"	"	3/8x4x28	10x16	10x10
	2x8	2x6	–	–	–	72	66	63	36	32	29	2x4	2x4	2x4	½x4x21	12x16	10x10
	2x10	4+4	–	–	–	100	92	92	50	46	30	"	"	"	½x4x27	14x16	12x10
	2x12	4+6	–	–	–	–	128	125	66	60	57	"	"	"	½x4x34	18x16	14x12
	2x12	6+6	–	–	–	–	–	–	63	57	54	"	"	"	½x4x34	"	16x12
5/12 Slope	2x4	2x4	64	61	59	28	25	24	14	0	0	2x4	2x4	2x4	3/8x3½x15	8x12	8x8
	2x6	2x4	100+	100+	100+	55	51	41	27	0	0	"	"	"	½x4x15	10x16	8x10
	2x6	2x6	–	–	–	53	50	48	26	23	21	"	"	"	3/8x4x26	"	"
	2x8	2x6	–	–	–	79	73	70	39	36	33	2x4	2x4	2x4	½x4x19	12x16	8x10
	2x10	4+4	–	–	–	100+	100+	100+	55	50	42	"	"	"	½x4x24	14x16	12x10
	2x12	4+6	–	–	–	–	–	–	72	65	62	"	"	"	½x4x30	18x16	12x12
	2x12	6+6	–	–	–	–	–	–	69	63	59	"	"	"	½x4x30	"	14x12

1600f Lumber

Slope	Top chord	Bottom chord	2' — 0	2' — 5	2' — 8	4' — 0	4' — 5	4' — 8	8' — 0	8' — 5	8' — 8	W1	W2	W3	A T H W	B H W	C H W
			Truss spacing, ft. — Ceiling dead load, psf — Max. snow + roof dead load, psf									Web member sizes			Gusset Sizes, in.		
3/12 Slope	2x4	2x4	61	58	56	26	23	22	13	0	0	2x4	2x4	2x4	3/8x3½x23	8x12	8x8
	2x6	2x4	100+	100+	100+	46	45	35	23	0	0	"	"	"	½x4x20	10x16	8x10
	2x6	2x6	–	–	–	49	45	44	24	21	18	"	"	"	3/8x4x36	"	8x12
	2x8	2x6	–	–	–	65	59	56	32	27	24	2x4	2x4	2x4	½x4x24	12x16	8x12
	2x10	4+4	–	–	–	100+	97	92	52	46	37	"	"	"	½x4x34	14x16	10x12
	2x12	6+4	–	–	–	–	100+	100+	64	57	54	"	"	"	½x4x40	16x16	12x12
	4x12	6+6	–	–	–	–	–	–	61	54	51	"	"	"	½x4x44	"	16x10
4/12 Slope	2x4	2x4	70	66	64	30	28	26	15	12	0	2x4	2x4	2x4	3/8x3½x20	8x12	8x10
	2x6	2x4	100+	100+	100+	60	56	48	30	15	0	"	"	"	½x4x19	10x16	10x10
	2x6	2x6	–	–	–	57	53	51	28	25	23	"	"	"	½x4x19	"	"
	2x8	2x6	–	–	–	87	80	77	43	40	37	2x4	2x4	2x4	½x4x25	14x16	10x12
	2x10	4+4	–	–	–	100+	100+	100+	60	55	49	"	"	"	½x4x32	16x16	12x12
	2x12	6+4	–	–	–	–	–	–	81	73	72	"	"	"	2–½x6x28	16x20	16x12
	2x12	6+6	–	–	–	–	–	–	76	69	66	"	"	"	½x4x40	"	18x12
5/12 Slope	2x4	2x4	76	73	71	33	31	29	16	14	0	2x4	2x4	2x4	3/8x3½x18	8x12	8x8
	2x6	2x4	100+	100+	100+	66	62	53	33	19	0	"	"	"	½x4x18	10x16	8x10
	2x6	2x6	–	–	–	64	60	57	32	28	27	"	"	"	½x4x17	"	"
	2x8	2x6	–	–	–	95	88	85	47	44	42	2x4	2x4	2x4	½x4x23	14x16	10x10
	2x10	4+4	–	–	–	100+	100+	100+	66	60	54	"	"	"	½x4x29	16x16	10x12
	2x12	4+6	–	–	–	–	–	–	86	79	75	"	"	"	½x4x36	16x20	12x12
	2x12	6+6	–	–	–	–	–	–	83	75	72	"	"	"	½x4x36	"	16x12

22' SPAN, 2-WEB

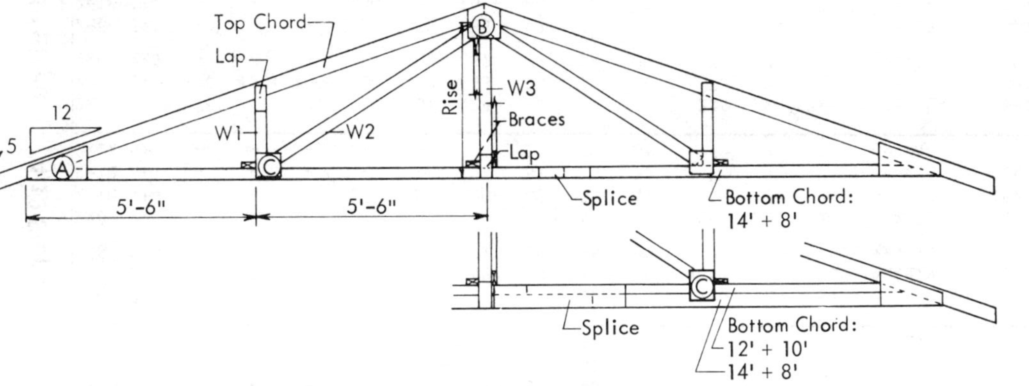

Gussets B and C are 3/8" thick plywood.

4+4, 4+6, 6+6 indicates stacked lower chord.
4&4, 6&4, indicate double web; a 2x4 is attached to the web member to increase its stiffness.

Before selecting heel gusset A, see pages 8 and 9.

Web Lengths

Roof Slope	Rise	Top Chord	W1	W2	W3
3/12	2'-9"	12'	2'	6'	3'
4/12	3'-8"	12'	2'	7'	4'
5/12	4'-7"	13'	2'	7'	5'

1100f Lumber

			Truss spacing, ft. 2'			4'			8'			Web member sizes			Gusset Sizes, in. A	B	C
	Top chord	Bottom chord	Ceiling dead load, psf 0	5	8	0	5	8	0	5	8	W1	W2	W3	T H W	H W	H W
			---Max. snow + roof dead load, psf---														
3/12 Slope	2x4	2x4	36	33	31	15	12	0	0	0	0	2x4	2x4	2x4	3/8x3½x16	8x12	8x8
	2x6	2x4	61	59	54	26	15	0	13	0	0	"	"	"	3/8x4x23	10x12	"
	2x6	2x6	67	62	59	29	25	23	14	0	0	"	"	"	3/8x4x26	"	"
	2x8	2x6	85	77	72	37	32	29	18	13	0	2x4	2x4	2x4	3/8x4x31	12x12	8x8
	2x10	4+4	100+	100+	100+	58	51	46	29	16	0	"	"	"	½x4x28	14x16	10x10
	2x12	4+6	–	–	–	73	66	62	36	31	24	"	"	"	3/8x4x47	16x16	12x10
	2x12	6+6	–	–	–	74	66	62	37	31	27	"	"	"	3/8x4x50	"	14x10
4/12 Slope	2x4	2x4	41	39	37	17	15	0	0	0	0	2x4	2x4	2x4	3/8x3½x14	8x12	8x8
	2x6	2x4	81	76	74	35	22	0	17	0	0	"	"	"	3/8x4x23	10x12	8x10
	2x6	2x6	78	73	70	34	30	28	17	13	0	"	"	"	3/8x4x23	"	"
	2x8	2x6	100+	100+	100+	51	47	45	25	21	0	2x4	2x4	2x4	½x4x18	14x12	8x10
	2x10	4+4	–	–	–	70	64	58	35	22	0	"	"	"	½x4x22	16x12	12x10
	2x12	4+6	–	–	–	92	83	82	46	40	32	"	"	"	½x4x28	16x16	14x10
	2x12	6+6	–	–	–	88	80	76	44	39	36	"	"	"	½x4x34	"	16x10
5/12 Slope	2x4	2x4	44	43	41	19	17	0	0	0	0	2x4	2x4	2x4	3/8x3½x12	8x12	8x8
	2x6	2x4	89	83	83	38	29	0	19	0	0	"	"	"	3/8x4x21	10x12	"
	2x6	2x6	86	81	78	37	34	32	18	16	0	"	"	"	3/8x4x21	10x16	8x10
	2x8	2x6	100+	100+	100+	56	52	50	28	24	13	2x4	2x4	2x4	½x4x16	12x16	8x10
	2x10	4+4	–	–	–	77	71	67	38	31	0	"	"	"	½x4x19	14x16	10x10
	2x12	4+6	–	–	–	100	91	92	50	45	38	"	"	"	½x4x24	16x16	12x10
	2x12	6+6	–	–	–	–	88	88	48	44	41	"	"	"	½x4x28	"	16x10

1400f Lumber

Slope	Top chord	Bottom chord	2' spacing, ceiling 0	2' spacing, ceiling 5	2' spacing, ceiling 8	4' spacing, ceiling 0	4' spacing, ceiling 5	4' spacing, ceiling 8	8' spacing, ceiling 0	8' spacing, ceiling 5	8' spacing, ceiling 8	Web W1	Web W2	Web W3	Gusset A T H W	Gusset B H W	Gusset C H W
			Max. snow + roof dead load, psf														
3/12 Slope	2x4	2x4	44	42	39	19	16	0	0	0	0	2x4	2x4	2x4	3/8x3½x19	8x12	8x8
	2x6	2x4	81	78	77	35	31	0	17	0	0	"	"	"	½x4x17	10x16	8x10
	2x6	2x6	82	76	73	35	32	29	17	14	12	"	"	"	3/8x4x31	"	"
	2x8	2x6	100+	100+	100+	48	44	40	24	19	14	2x4	2x4	2x4	½x4x23	12x16	8x10
	2x10	4+4	-	-	-	76	68	65	38	31	0	"	"	"	½x4x29	14x16	10x12
	2x12	4+6	-	-	-	95	85	85	47	42	38	"	"	"	½x4x36	16x16	12x12
	2x12	6+6	-	-	-	91	81	81	45	39	36	"	"	"	½x4x44	"	14x12
4/12 Slope	2x4	2x4	51	48	47	22	19	14	0	0	0	2x4	2x4	2x4	3/8x3½x17	8x12	8x8
	2x6	2x4	100	93	93	43	40	14	21	0	0	"	"	"	½x4x16	12x12	8x10
	2x6	2x6	96	89	86	41	39	36	20	17	15	"	"	"	3/8x4x28	10x16	10x10
	2x8	2x6	100+	100+	100+	62	58	55	31	27	22	2x4	2x4	2x4	½x4x22	12x16	10x10
	2x10	4+4	-	-	-	86	79	78	43	38	16	"	"	"	½x4x26	14x16	12x10
	2x12	4+6	-	-	-	100+	100+	100+	56	51	49	"	"	"	½x4x33	18x16	14x12
	2x12	6+6	-	-	-	-	98	99	54	49	46	"	"	"	½x4x34	16x20	16x12
5/12 Slope	2x4	2x4	55	53	52	24	22	20	12	0	0	2x4	2x4	2x4	3/8x3½x15	8x12	8x8
	2x6	2x4	100+	100+	100+	47	45	29	23	0	0	"	"	"	½x4x14	10x16	8x10
	2x6	2x6	-	-	-	46	43	42	23	20	18	"	"	"	3/8x4x25	"	"
	2x8	2x6	-	-	-	69	64	61	34	31	29	2x4	2x4	2x4	½x4x19	12x16	8x10
	2x10	4+4	-	-	-	95	87	87	47	43	22	"	"	"	½x4x24	16x16	12x10
	2x12	4+6	-	-	-	100+	100+	100+	61	56	53	"	"	"	½x4x29	18x16	12x12
	2x12	6+6	-	-	-	-	-	-	59	54	51	"	"	"	½x4x29	16x20	14x12

1600f Lumber

Slope	Top chord	Bottom chord	2' spacing, ceiling 0	2' spacing, ceiling 5	2' spacing, ceiling 8	4' spacing, ceiling 0	4' spacing, ceiling 5	4' spacing, ceiling 8	8' spacing, ceiling 0	8' spacing, ceiling 5	8' spacing, ceiling 8	Web W1	Web W2	Web W3	Gusset A T H W	Gusset B H W	Gusset C H W
			Max. snow + roof dead load, psf														
3/12 Slope	2x4	2x4	53	50	48	23	20	18	0	0	0	2x4	2x4	2x4	3/8x3½x23	8x12	8x8
	2x6	2x4	97	93	94	42	40	20	21	0	0	"	"	"	½x4x20	10x16	8x10
	2x6	2x6	99	92	88	43	40	37	21	18	15	"	"	"	½x4x22	"	8x12
	2x8	2x6	-	100+	100+	58	52	49	29	24	20	2x4	2x4	2x4	½x4x27	12x16	10x10
	2x10	4+4	-	-	-	92	81	80	46	39	20	"	"	"	½x4x34	14x16	12x10
	2x12	4+6	-	-	-	100+	100+	100+	57	51	48	"	"	"	½x4x41	16x20	14x10
	2x12	6+6	-	-	-	-	-	-	55	49	46	"	"	"	½x4x44	"	16x10
4/12 Slope	2x4	2x4	60	58	56	26	24	22	13	0	0	2x4	2x4	2x4	3/8x3½x20	8x12	8x10
	2x6	2x4	100+	100+	100+	52	49	29	26	0	0	"	"	"	½x4x19	10x16	10x10
	2x6	2x6	-	-	-	50	47	45	25	21	19	"	"	"	½x4x18	"	"
	2x8	2x6	-	-	-	75	70	67	37	33	31	2x4	2x4	2x4	½x4x24	14x16	10x12
	2x10	4+4	-	-	-	100+	96	96	52	48	30	"	"	"	½x4x31	14x20	12x12
	2x12	4+6	-	-	-	-	100+	100+	68	62	59	"	"	"	½x4x39	16x20	16x12
	2x12	6+6	-	-	-	-	-	-	65	59	56	"	"	"	½x4x39	"	18x12
5/12 Slope	2x4	2x4	66	63	61	28	26	25	14	0	0	2x4	2x4	2x4	3/8x3½x17	8x12	8x8
	2x6	2x4	100+	100+	100+	57	54	41	28	17	0	"	"	"	½x4x17	10x16	8x10
	2x6	2x6	-	-	-	55	52	50	27	24	23	"	"	"	½x4x17	"	"
	2x8	2x6	-	-	-	83	77	74	41	37	35	2x4	2x4	2x4	½x4x22	14x16	10x10
	2x10	4+4	-	-	-	100+	100+	100+	57	52	43	"	"	"	½x4x28	14x20	10x12
	2x12	4+6	-	-	-	-	-	-	74	68	64	"	"	"	½x4x35	16x20	14x12
	2x12	6+6	-	-	-	-	-	-	71	65	62	"	"	"	½x4x35	18x20	16x12

24' SPAN, 2-WEB

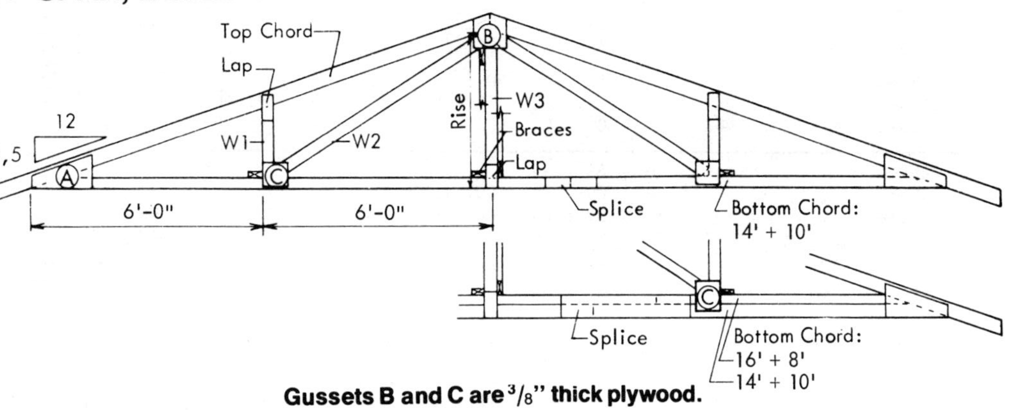

Gussets B and C are ³/₈" thick plywood.

4+4, 4+6, 6+6 indicates stacked lower chord.
4&4, 6&4, indicate double web; a 2x4 is attached to the web member to increase its stiffness.

Before selecting heel gusset A, see pages 8 and 9.

Web Lengths

Roof Slope	Rise	Top Chord	W1	W2	W3
3/12	3'-0"	13'	2'	7'	3'
4/12	4'-0"	13'	2'	7'	4'
5/12	5'-0"	14'	3'	8'	5'

1100f Lumber

			Truss spacing, ft.									Web member sizes			Gusset Sizes, in.		
			2'			4'			8'						A	B	C
			Ceiling dead load, psf														
	Top chord	Bottom chord	0	5	8	0	5	8	0	5	8	W1	W2	W3	T H W	H W	H W
			---Max. snow + roof dead load, psf---														
3/12 Slope	2x4	2x4	31	29	27	13	0	0	0	0	0	2x4	2x4	2x4	3/8x3½x16	8x12	8x8
	2x6	2x4	56	53	46	24	0	0	12	0	0	"	"	"	3/8x4x23	10x12	"
	2x6	2x6	59	55	52	25	22	19	12	0	0	"	"	"	3/8x4x26	"	"
	2x8	2x6	76	69	65	33	28	25	16	0	0	2x4	2x4	2x4	3/8x4x31	12x12	8x8
	2x10	4+4	100+	100+	100+	52	46	34	26	0	0	"	"	"	½x4x28	14x16	10x10
	2x12	4+6	-	-	-	66	58	54	33	27	16	"	"	"	3/8x4x47	16x16	12x10
	2x12	6+6	-	-	-	67	61	57	33	28	24	"	"	"	3/8x4x51	"	14x12
4/12 Slope	2x4	2x4	36	33	32	15	13	0	0	0	0	2x4	2x4	2x4	3/8x3½x14	8x12	8x8
	2x6	2x4	71	67	54	31	13	0	15	0	0	"	"	"	3/8x4x22	10x12	8x10
	2x6	2x6	68	64	62	30	26	25	15	0	0	"	"	"	3/8x4x23	"	"
	2x8	2x6	100+	95	91	44	41	38	22	17	0	2x4	2x4	2x4	½x4x18	14x12	8x10
	2x10	4+4	-	100+	100+	61	56	46	30	19	0	"	"	"	½x4x23	14x16	12x10
	2x12	4+6	-	-	-	80	73	72	40	35	23	"	"	"	½x4x28	16x16	14x10
	2x12	6+6	-	-	-	76	70	66	38	33	31	"	"	"	½x4x34	"	16x10
5/12 Slope	2x4	2x4	39	37	35	17	15	0	0	0	0	2x4	2x4	2x4	3/8x3½x12	8x12	8x8
	2x6	2x4	78	74	71	34	17	0	17	0	0	"	"	"	3/8x4x20	10x12	"
	2x6	2x6	76	72	69	33	30	28	16	13	0	"	"	"	3/8x4x20	"	"
	2x8	2x6	100+	100+	100+	49	46	44	24	20	0	2x4	2x4	2x4	½x4x16	12x16	8x10
	2x10	4+4	-	-	-	68	62	53	34	18	0	"	"	"	½x4x18	14x16	10x10
	2x12	4+6	-	-	-	87	80	79	43	39	27	"	"	"	½x4x23	16x16	12x10
	2x12	6+6	-	-	-	84	77	73	42	37	35	"	"	"	½x4x28	18x16	16x10

1400f Lumber

Slope	Top chord	Bottom chord	2' — 0	2' — 5	2' — 8	4' — 0	4' — 5	4' — 8	8' — 0	8' — 5	8' — 8	W1	W2	W3	A T H W	B H W	C H W
			Truss spacing, ft. / Ceiling dead load, psf / ---Max. snow + roof dead load, psf---									Web member sizes			Gusset Sizes, in.		
3/12 Slope	2x4	2x4	39	37	34	17	14	0	0	0	0	2x4	2x4	2x4	3/8x3½x19	8x12	8x8
	2x6	2x4	74	70	68	32	22	14	16	0	0	"	"	"	½x4x17	10x16	8x10
	2x6	2x6	72	67	64	31	28	26	15	12	0	"	"	"	3/8x4x31	"	"
	2x8	2x6	100	90	91	43	38	35	21	16	0	2x4	2x4	2x4	½x4x23	12x16	8x10
	2x10	4+4	–	100+	100+	66	60	54	33	22	0	"	"	"	½x4x32	14x16	10x12
	2x12	4+6	–	–	–	86	76	75	43	37	31	"	"	"	½x4x36	16x16	12x12
	2x12	6+6	–	–	–	82	74	70	41	35	32	"	"	"	½x4x44	"	16x10
4/12 Slope	2x4	2x4	44	42	41	19	17	0	0	0	0	2x4	2x4	2x4	3/8x3½x17	8x12	8x8
	2x6	2x4	87	82	81	38	30	0	19	0	0	"	"	"	½x4x16	12x12	8x10
	2x6	2x6	85	79	76	37	33	31	18	15	13	"	"	"	3/8x4x28	10x16	10x10
	2x8	2x6	100+	100+	100+	55	51	49	27	23	16	2x4	2x4	2x4	½x4x22	12x16	10x10
	2x10	4+4	–	–	–	75	69	68	37	32	0	"	"	"	½x4x25	14x16	12x10
	2x12	4+6	–	–	–	98	90	90	49	45	41	"	"	"	½x4x32	18x16	14x12
	2x12	6+6	–	–	–	94	86	86	47	43	39	"	"	"	½x4x34	16x20	16x12
5/12 Slope	2x4	2x4	48	46	45	21	19	0	0	0	0	2x4	2x4	2x4	3/8x3½x14	8x12	8x8
	2x6	2x4	96	91	91	42	39	0	21	0	0	"	"	"	½x4x14	10x16	8x10
	2x6	2x6	94	88	86	41	38	36	20	17	15	"	"	"	3/8x4x25	"	"
	2x8	2x6	100+	100+	100+	61	56	54	30	27	19	2x4	2x4	2x4	½x4x20	12x16	8x10
	2x10	4+4	–	–	–	83	77	76	41	37	0	"	"	"	½x4x23	16x16	12x10
	2x12	4+6	–	–	–	100+	98	100	53	49	47	"	"	"	½x4x29	16x20	12x12
	2x12	6+6	–	–	–	–	95	96	52	47	45	"	"	"	½x4x29	"	14x12

1600f Lumber

Slope	Top chord	Bottom chord	2' — 0	2' — 5	2' — 8	4' — 0	4' — 5	4' — 8	8' — 0	8' — 5	8' — 8	W1	W2	W3	A T H W	B H W	C H W
			Truss spacing, ft. / Ceiling dead load, psf / ---Max. snow + roof dead load, psf---									Web member sizes			Gusset Sizes, in.		
3/12 Slope	2x4	2x4	47	44	43	20	17	0	0	0	0	2x4	2x4	2x4	3/8x3½x22	8x12	8x8
	2x6	2x4	88	84	83	38	33	0	19	0	0	"	"	"	½x4x20	10x16	8x10
	2x6	2x6	87	81	78	38	34	32	19	15	13	"	"	"	3/8x4x36	"	10x10
	2x8	2x6	100+	100+	100+	52	47	44	26	21	16	2x4	2x4	2x4	½x4x27	12x16	10x10
	2x10	4+4	–	–	–	80	73	71	40	34	0	"	"	"	½x4x33	14x20	12x10
	2x12	4+6	–	–	–	100+	96	92	51	45	41	"	"	"	½x4x41	16x20	14x12
	2x12	6+6	–	–	–	–	90	91	50	45	41	"	"	"	½x4x44	"	16x12
4/12 Slope	2x4	2x4	53	50	49	23	21	14	0	0	0	2x4	2x4	2x4	3/8x3½x19	8x12	8x10
	2x6	2x4	100+	99	99	45	43	14	22	0	0	"	"	"	½x4x18	10x16	10x10
	2x6	2x6	–	95	92	44	41	39	22	19	17	"	"	"	½x4x18	"	"
	2x8	2x6	–	100+	100+	66	61	59	33	29	26	2x4	2x4	2x4	½x4x26	14x16	10x12
	2x10	4+4	–	–	–	91	84	83	45	42	18	"	"	"	½x4x31	14x20	12x12
	2x12	4+6	–	–	–	100+	100+	100+	59	54	51	"	"	"	½x4x38	16x20	16x12
	2x12	6+6	–	–	–	–	–	–	56	51	49	"	"	"	½x4x38	18x20	18x14
5/12 Slope	2x4	2x4	57	55	54	25	23	21	12	0	0	2x4	2x4	2x4	3/8x3½x17	8x12	8x8
	2x6	2x4	100+	100+	100+	50	47	21	25	0	0	"	"	"	½x4x17	10x16	8x10
	2x6	2x6	–	–	–	49	46	44	24	21	20	"	"	"	½x4x17	"	"
	2x8	2x6	–	–	–	73	68	65	36	33	31	2x4	2x4	2x4	½x4x22	14x16	8x12
	2x10	4+4	–	–	–	100	92	93	50	46	22	"	"	"	½x4x28	14x20	10x12
	2x12	4+6	–	–	–	–	100+	100+	64	59	56	"	"	"	½x4x34	18x20	14x12
	2x12	6+6	–	–	–	–	–	–	62	57	54	"	"	"	½x4x34	"	16x12

26' SPAN, 2-WEB

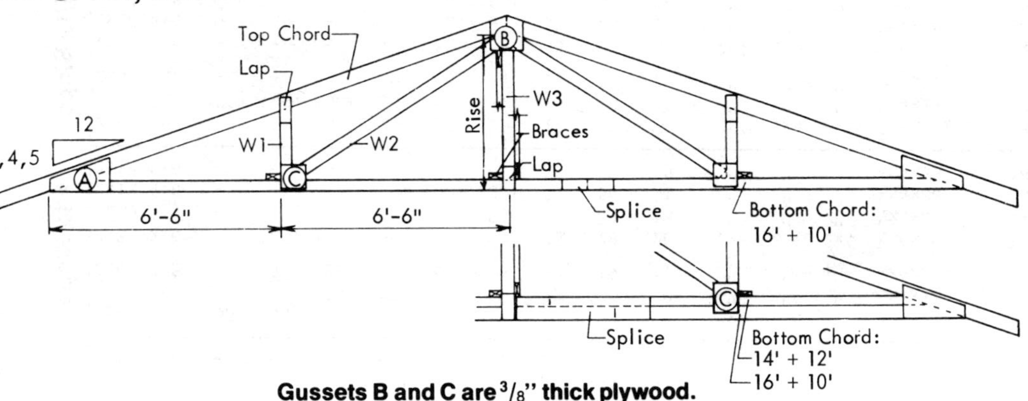

Gussets B and C are $^3/_8$" thick plywood.

4+4, 4+6, 6+6 indicates stacked lower chord.
4&4, 6&4, indicate double web; a 2x4 is attached to the web member to increase its stiffness.

Before selecting heel gusset A, see pages 8 and 9 .

Web Lengths

Roof Slope	Rise	Top Chord	W1	W2	W3
3/12	3'-3"	14'	2'	7'	3'
4/12	4'-4"	15'	3'	8'	4'
5/12	5'-5"	15'	3'	8'	6'

1100f Lumber

			Truss spacing, ft.												Gusset Sizes, in.		
			2'			4'			8'			Web member sizes			A	B	C
			Ceiling dead load, psf														
	Top chord	Bottom chord	0	5	8	0	5	8	0	5	8	W1	W2	W3	T H W	H W	H W
			---Max. snow + roof dead load, psf---														
3/12 Slope	2x4	2x4	28	25	24	12	0	0	0	0	0	2x4	2x4	2x4	3/8x3½x16	8x12	8x8
	2x6	2x4	52	48	33	22	0	0	0	0	0	"	"	"	3/8x4x23	10x12	"
	2x6	2x6	52	49	47	23	19	17	0	0	0	"	"	"	3/8x4x26	"	"
	2x8	2x6	69	62	59	30	25	22	15	0	0	2x4	2x4	2x4	3/8x4x31	12x16	8x10
	2x10	4+4	100+	100+	97	47	41	21	23	0	0	"	"	"	½x4x28	14x16	10x10
	2x12	4+6	–	–	100+	59	52	49	29	23	0	"	"	"	½x4x36	16x16	12x10
	2x12	6+6	–	–	–	59	53	51	29	25	21	"	"	"	3/8x4x50	"	14x12
4/12 Slope	2x4	2x4	31	29	28	13	0	0	0	0	0	2x4	2x4	2x4	3/8x3½x13	8x12	8x8
	2x6	2x4	63	59	46	27	0	0	13	0	0	"	"	"	3/8x4x22	10x12	8x10
	2x6	2x6	61	57	55	26	23	22	13	0	0	"	"	"	3/8x4x23	"	"
	2x8	2x6	91	85	81	39	36	32	19	12	0	2x4	2x4	2x4	3/8x4x30	14x12	8x10
	2x10	4+4	100+	100+	100+	54	50	31	27	0	0	"	"	"	½x4x23	14x16	12x10
	2x12	4+6	–	–	–	70	64	62	35	29	15	"	"	"	½x4x28	16x16	14x10
	2x12	6+6	–	–	–	67	62	59	33	29	27	"	"	"	3/8x4x46	18x16	16x12
5/12 Slope	2x4	2x4	34	33	31	14	0	0	0	0	0	2x4	2x4	2x4	3/8x3½x12	8x12	8x8
	2x6	2x4	69	65	52	30	0	0	15	0	0	"	"	"	3/8x4x20	10x12	"
	2x6	2x6	67	64	62	29	26	25	14	12	0	"	"	"	3/8x4x20	"	"
	2x8	2x6	100+	94	91	44	41	38	22	15	0	2x4	2x4	2x4	½x4x15	12x16	8x10
	2x10	4+4	–	100+	100+	60	56	42	30	0	0	"	"	"	½x4x21	14x16	10x10
	2x12	4+6	–	–	–	77	71	70	38	34	19	"	"	"	½x4x23	16x16	14x10
	2x12	6+6	–	–	–	75	69	65	37	33	30	"	"	"	½x4x28	18x16	16x10

1400f Lumber

Slope	Top chord	Bottom chord	2' (Ceiling dead load, psf) 0	2' 5	2' 8	4' 0	4' 5	4' 8	8' 0	8' 5	8' 8	W1	W2	W3	A T H W	B H W	C H W
			---Max. snow + roof dead load, psf---									Web member sizes			Gusset Sizes, in.		
3/12 Slope	2x4	2x4	34	32	30	15	12	0	0	0	0	2x4	2x4	2x4	3/8x3½x19	8x12	8x8
	2x6	2x4	68	63	54	29	13	0	14	0	0	"	"	"	½x4x17	10x16	8x10
	2x6	2x6	65	60	58	28	24	23	14	0	0	"	"	"	3/8x4x31	"	"
	2x8	2x6	90	82	82	39	34	31	19	14	0	2x4	2x4	2x4	½x4x23	12x16	8x12
	2x10	4+4	100+	100+	100+	58	53	46	29	14	0	"	"	"	½x4x31	14x16	12x10
	2x12	4+6	–	–	–	77	69	64	38	32	23	"	"	"	½x4x36	16x16	14x10
	2x12	6+6	–	–	–	72	66	62	36	31	28	"	"	"	½x4x44	16x20	16x12
4/12 Slope	2x4	2x4	39	37	35	17	15	0	0	0	0	2x4	2x4	2x4	3/8x3½x16	8x12	8x8
	2x6	2x4	78	73	70	34	20	0	17	0	0	"	"	"	½x4x15	12x12	8x10
	2x6	2x6	75	71	68	33	29	28	16	13	0	"	"	"	3/8x4x27	"	10x10
	2x8	2x6	100+	100+	100+	49	45	43	24	21	0	2x4	2x4	2x4	½x4x21	12x16	10x10
	2x10	4+4	–	–	–	67	62	57	33	21	0	"	"	"	½x4x24	14x16	12x10
	2x12	4+6	–	–	–	87	79	75	43	38	31	"	"	"	½x4x31	18x16	14x12
	2x12	6+6	–	–	–	83	76	72	41	37	34	"	"	"	½x4x34	16x20	16x12
5/12 Slope	2x4	2x4	42	41	40	18	16	0	0	0	0	2x4	2x4	2x4	3/8x3½x14	8x12	8x8
	2x6	2x4	85	81	80	37	26	15	18	0	0	"	"	"	½x4x14	10x16	8x10
	2x6	2x6	83	79	76	36	33	32	18	15	0	"	"	"	3/8x4x25	"	"
	2x8	2x6	100+	100+	100+	54	50	48	27	23	0	2x4	2x4	2x4	½x4x19	12x16	8x10
	2x10	4+4	–	–	–	74	69	67	37	29	15	"	"	"	½x4x22	16x16	12x10
	2x12	4+6	–	–	–	95	87	88	47	43	38	"	"	"	½x4x28	16x20	12x12
	2x12	6+6	–	–	–	92	85	81	46	42	39	"	"	"	½x4x28	"	14x12

1600f Lumber

Slope	Top chord	Bottom chord	2' (Ceiling dead load, psf) 0	2' 5	2' 8	4' 0	4' 5	4' 8	8' 0	8' 5	8' 8	W1	W2	W3	A T H W	B H W	C H W
			---Max. snow + roof dead load, psf---									Web member sizes			Gusset Sizes, in.		
3/12 Slope	2x4	2x4	41	39	37	18	15	0	0	0	0	2x4	2x4	2x4	3/8x3½x22	8x12	8x8
	2x6	2x4	82	76	73	35	23	0	17	0	0	"	"	"	½x4x20	10x16	8x10
	2x6	2x6	78	73	70	34	30	28	17	13	0	"	"	"	3/8x4x36	"	8x12
	2x8	2x6	100+	98	99	47	42	39	23	18	0	2x4	2x4	2x4	½x4x27	12x16	10x10
	2x10	4+4	–	–	–	71	65	60	35	24	0	"	"	"	½x4x33	14x20	12x10
	2x12	4+6	–	–	–	93	82	81	46	40	35	"	"	"	½x4x41	16x20	14x12
	2x12	6+6	–	–	–	88	80	75	44	39	36	"	"	"	½x4x44	"	16x12
4/12 Slope	2x4	2x4	46	44	43	20	18	0	0	0	0	2x4	2x4	2x4	3/8x3½x19	8x12	8x10
	2x6	2x4	93	88	87	40	34	0	20	0	0	"	"	"	½x4x18	10x16	10x10
	2x6	2x6	90	85	82	39	37	34	19	16	14	"	"	"	½x4x18	"	"
	2x8	2x6	100+	100+	100+	59	55	52	29	26	15	2x4	2x4	2x4	½x4x25	14x16	10x12
	2x10	4+4	–	–	–	81	74	73	40	35	0	"	"	"	½x4x29	14x20	12x12
	2x12	4+6	–	–	–	100+	96	97	52	48	44	"	"	"	½x4x37	18x20	16x14
	2x12	6+6	–	–	–	100	92	92	50	46	43	"	"	"	½x4x38	"	18x14
5/12 Slope	2x4	2x4	50	48	47	22	20	0	0	0	0	2x4	2x4	2x4	3/8x3½x16	8x12	8x8
	2x6	2x4	100+	97	97	44	42	0	22	0	0	"	"	"	½x4x16	10x16	8x10
	2x6	2x6	–	94	91	43	41	39	21	19	17	"	"	"	½x4x17	"	"
	2x8	2x6	–	100+	100+	65	60	58	32	29	21	2x4	2x4	2x4	½x4x23	14x16	10x10
	2x10	4+4	–	–	–	89	83	82	44	41	0	"	"	"	½x4x28	14x20	10x12
	2x12	4+6	–	–	–	100+	100+	100+	57	52	50	"	"	"	½x4x34	18x20	14x12
	2x12	6+6	–	–	–	–	–	–	55	51	48	"	"	"	½x4x34	"	16x12

28' SPAN, 2-WEB

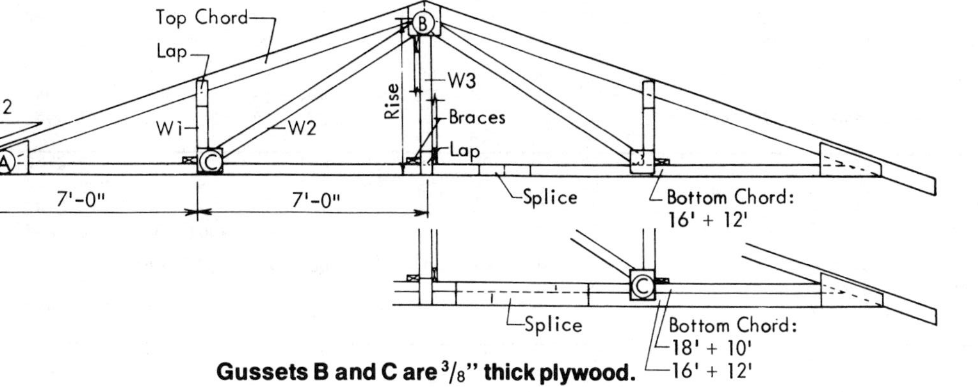

Gussets B and C are $^3/_8$" thick plywood.

4+4, 4+6, 6+6 indicates stacked lower chord.
4&4, 6&4, indicate double web; a 2x4 is attached to the web member to increase its stiffness.

Before selecting heel gusset A, see pages 8 and 9.

Web Lengths

Roof Slope	Rise	Top Chord	W1	W2	W3
3/12	3'-6"	15'	2'	8'	4'
4/12	4'-8"	16'	2'	8'	5'
5/12	5'-10"	16'	3'	9'	6'

1100f Lumber

Max. snow + roof dead load, psf, by truss spacing (2', 4', 8') and ceiling dead load (0, 5, 8 psf). Web member sizes W1, W2, W3. Gusset sizes, in.: A (T H W), B (H W), C (H W).

Slope	Top chord	Bottom chord	2': 0	2': 5	2': 8	4': 0	4': 5	4': 8	8': 0	8': 5	8': 8	W1	W2	W3	A T H W	B H W	C H W
3/12 Slope	2x4	2x4	25	22	21	0	0	0	0	0	0	2x4	2x4	2x4	3/8x3½x15	8x12	8x8
	2x6	2x4	48	44	21	21	0	0	0	0	0	"	"	"	3/8x4x24	10x12	"
	2x6	2x6	47	44	42	20	17	15	0	0	0	"	"	"	3/8x4x26	10x16	8x10
	2x8	2x6	63	57	54	27	22	18	13	0	0	2x4	4&4	2x4	3/8x4x31	12x16	8x10
	2x10	4+4	98	88	87	42	35	0	21	0	0	"	"	"	½x4x28	14x16	10x10
	2x12	4+6	-	100+	100+	54	47	43	27	18	0	"	"	"	½x4x36	16x16	12x12
	2x12	6+6	-	-	-	52	48	45	26	21	14	"	"	"	3/8x4x49	"	14x12
4/12 Slope	2x4	2x4	28	26	25	12	0	0	0	0	0	2x4	2x4	2x4	3/8x3½x13	8x12	8x8
	2x6	2x4	56	53	30	24	0	0	0	0	0	"	"	"	3/8x4x22	10x12	8x10
	2x6	2x6	55	52	50	24	21	19	12	0	0	"	4&4	"	3/8x4x22	"	"
	2x8	2x6	82	76	73	35	32	25	17	0	0	2x4	4&4	2x4	3/8x4x30	14x12	10x10
	2x10	4+4	100+	100+	100+	49	45	18	24	0	0	"	"	"	½x4x22	14x16	12x10
	2x12	4+6	-	-	-	63	58	52	31	24	0	"	"	"	½x4x28	16x16	14x10
	2x12	6+6	-	-	-	60	55	53	30	26	21	"	"	"	3/8x4x45	18x16	16x12
5/12 Slope	2x4	2x4	30	28	27	13	0	0	0	0	0	2x4	2x4	2x4	3/8x3½x12	8x12	8x8
	2x6	2x4	62	59	42	27	0	0	13	0	0	"	4&4	"	3/8x4x19	10x12	"
	2x6	2x6	60	57	55	26	23	22	13	0	0	"	"	"	3/8x4x20	"	"
	2x8	2x6	90	85	82	39	36	31	19	0	0	2x4	4&4	2x4	½x4x15	12x16	8x10
	2x10	4+4	100+	100+	100+	54	50	24	27	0	0	"	"	"	½x4x20	14x16	10x10
	2x12	4+6	-	-	-	69	64	62	34	28	0	"	"	"	½x4x22	16x16	12x10
	2x12	6+6	-	-	-	67	62	59	33	29	26	"	"	"	½x4x28	18x16	14x12

1400f Lumber

Truss spacing, ft. / Ceiling dead load, psf / ---Max. snow + roof dead load, psf---

Slope	Top chord	Bottom chord	2' 0	2' 5	2' 8	4' 0	4' 5	4' 8	8' 0	8' 5	8' 8	Web W1	Web W2	Web W3	Gusset A T H W	Gusset B H W	Gusset C H W
3/12 Slope	2x4	2x4	31	28	27	13	0	0	0	0	0	2x4	2x4	2x4	3/8x3½x18	8x12	8x8
	2x6	2x4	61	57	48	26	0	0	0	0	0	"	"	"	½x4x17	10x16	8x10
	2x6	2x6	58	54	52	25	22	20	12	0	0	"	"	"	3/8x4x30	"	"
	2x8	2x6	82	74	70	36	31	28	18	12	0	2x4	2x4	2x4	½x4x22	12x16	8x12
	2x10	4+4	100+	100+	100+	52	48	33	26	0	0	"	"	"	½x4x30	14x16	12x10
	2x12	4+6	–	–	–	68	62	58	34	29	14	"	"	"	½x4x36	16x16	14x10
	2x12	6+6	–	–	–	64	58	55	32	27	24	"	"	"	½x4x44	16x20	16x12
4/12 Slope	2x4	2x4	35	33	31	15	12	0	0	0	0	2x4	2x4	2x4	3/8x3½x16	8x12	8x8
	2x6	2x4	70	66	58	30	12	0	15	0	0	"	"	"	½x4x15	10x12	8x10
	2x6	2x6	68	64	62	29	26	25	14	0	0	"	"	"	3/8x4x27	12x12	10x10
	2x8	2x6	100+	94	90	44	41	38	22	18	0	2x4	2x4	2x4	½x4x20	12x16	10x10
	2x10	4+4	–	100+	100+	60	55	48	30	13	0	"	"	"	½x4x27	16x16	12x12
	2x12	4+6	–	–	–	77	71	67	38	34	21	"	"	"	½x4x30	16x20	14x12
	2x12	6+6	–	–	–	74	68	65	37	33	30	"	"	"	½x4x34	"	18x12
5/12 Slope	2x4	2x4	37	36	35	16	14	0	0	0	0	2x4	2x4	2x4	3/8x3½x14	8x12	8x8
	2x6	2x4	76	73	72	33	16	0	16	0	0	"	"	"	½x4x14	10x16	8x10
	2x6	2x6	75	71	69	32	30	28	16	13	0	"	"	"	3/8x4x24	"	"
	2x8	2x6	100+	100+	100+	48	45	44	24	21	0	2x4	2x4	2x4	½x4x18	12x16	8x10
	2x10	4+4	–	–	–	67	62	53	33	14	0	"	"	"	½x4x22	14x16	12x10
	2x12	4+6	–	–	–	85	79	75	42	38	27	"	"	"	½x4x28	16x20	12x12
	2x12	6+6	–	–	–	83	76	73	41	37	35	"	"	"	½x4x28	"	16x12

1600f Lumber

Truss spacing, ft. / Ceiling dead load, psf / ---Max. snow + roof dead load, psf---

Slope	Top chord	Bottom chord	2' 0	2' 5	2' 8	4' 0	4' 5	4' 8	8' 0	8' 5	8' 8	Web W1	Web W2	Web W3	Gusset A T H W	Gusset B H W	Gusset C H W
3/12 Slope	2x4	2x4	37	35	33	16	14	0	0	0	0	2x4	2x4	2x4	3/8x3½x21	8x12	8x8
	2x6	2x4	73	69	61	32	16	0	16	0	0	"	"	"	½x4x20	10x16	8x10
	2x6	2x6	70	65	63	30	27	25	15	12	0	"	"	"	3/8x4x36	"	8x12
	2x8	2x6	99	89	90	43	38	35	21	16	0	2x4	2x4	2x4	½x4x27	12x16	10x10
	2x10	4+4	–	100+	100+	63	58	50	31	16	0	"	"	"	½x4x37	14x20	12x12
	2x12	4+6	–	–	–	82	74	73	41	35	27	"	"	"	½x4x40	16x20	14x12
	2x12	6+6	–	–	–	78	71	67	39	34	31	"	"	"	½x4x44	18x24	16x12
4/12 Slope	2x4	2x4	41	40	39	18	16	0	0	0	0	2x4	2x4	2x4	3/8x3½x19	8x12	8x10
	2x6	2x4	83	79	75	36	23	0	18	0	0	"	"	"	½x4x17	10x16	10x10
	2x6	2x6	81	76	74	35	32	30	17	15	0	"	"	"	½x4x18	"	"
	2x8	2x6	100+	100+	100+	53	49	47	26	23	0	2x4	2x4	2x4	½x4x25	14x16	10x12
	2x10	4+4	–	–	–	72	67	61	36	23	0	"	"	"	½x4x29	14x20	12x12
	2x12	4+6	–	–	–	93	85	81	46	42	35	"	"	"	½x4x37	16x20	16x12
	2x12	6+6	–	–	–	90	82	78	45	41	38	"	"	"	½x4x37	18x20	18x14
5/12 Slope	2x4	2x4	44	43	42	19	17	0	0	0	0	2x4	2x4	2x4	3/8x3½x16	8x12	8x8
	2x6	2x4	91	87	86	39	28	0	19	0	0	"	"	"	½x4x16	10x16	8x10
	2x6	2x6	89	85	82	39	37	34	19	17	14	"	"	"	½x4x16	"	"
	2x8	2x6	100+	100+	100+	58	54	52	29	26	0	2x4	2x4	2x4	½x4x22	14x16	10x10
	2x10	4+4	–	–	–	80	74	71	40	30	0	"	"	"	½x4x26	14x20	10x12
	2x12	4+6	–	–	–	100+	94	95	51	47	42	"	"	"	½x4x33	18x20	14x12
	2x12	6+6	–	–	–	–	92	92	50	46	44	"	"	"	½x4x34	"	16x12

30' SPAN, 2-WEB

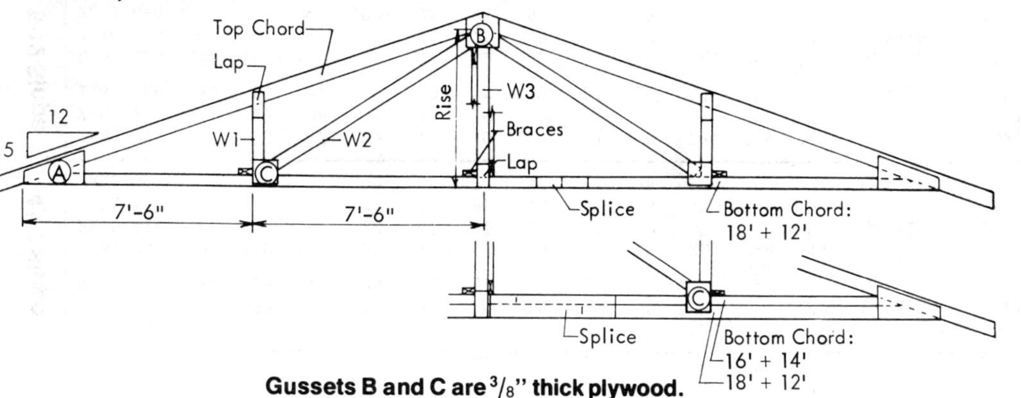

Gussets B and C are 3/8" thick plywood.

4+4, 4+6, 6+6 indicates stacked lower chord.
4&4, 6&4, indicate double web; a 2x4 is attached to the web member to increase its stiffness.

Before selecting heel gusset A, see pages 8 and 9.

Web Lengths

Roof Slope	Rise	Top Chord	W1	W2	W3
3/12	3'-9"	16'	2'	8'	4'
4/12	5'-0"	17'	3'	9'	5'
5/12	6'-3"	17'	3'	10'	6'

1100f Lumber

			Truss spacing, ft. — 2'			4'			8'			Web member sizes			Gusset Sizes, in.		
	Top chord	Bottom chord	Ceiling dead load, psf 0	5	8	0	5	8	0	5	8	W1	W2	W3	A T H W	B H W	C H W
			---Max. snow + roof dead load, psf---														
3/12 Slope	2x4	2x4	22	20	0	0	0	0	0	0	0	2x4	2x4	2x4	3/8x3½x14	8x12	8x8
	2x6	2x4	44	37	0	19	0	0	0	0	0	"	"	"	3/8x4x23	10x12	"
	2x6	2x6	43	40	37	18	15	12	0	0	0	"	"	"	3/8x4x25	10x16	8x10
	2x8	2x6	58	52	49	25	20	12	0	0	0	2x4	2x4	2x4	3/8x4x31	12x16	8x10
	2x10	4+4	88	81	78	38	28	0	19	0	0	"	4&4	"	½x4x28	14x16	10x10
	2x12	4+6	100+	100+	100+	50	43	36	25	13	0	"	"	"	½x4x36	16x16	12x12
	2x12	6+6	-	-	-	47	43	39	23	18	0	"	"	"	3/8x4x48	"	14x12
4/12 Slope	2x4	2x4	25	23	17	0	0	0	0	0	0	2x4	2x4	2x4	3/8x3½x13	8x12	8x8
	2x6	2x4	51	48	17	22	0	0	0	0	0	"	"	"	3/8x4x21	10x12	8x10
	2x6	2x6	49	47	45	21	19	17	0	0	0	"	"	"	3/8x4x22	"	"
	2x8	2x6	74	69	67	32	29	19	0	0	0	2x4	2x4	2x4	3/8x4x29	14x12	10x10
	2x10	4+4	100+	94	94	44	38	0	22	0	0	"	4&4	"	½x4x22	14x16	12x10
	2x12	4+6	-	100+	100+	57	52	46	28	18	0	"	"	"	½x4x28	18x16	14x12
	2x12	6+6	-	-	-	55	50	48	27	23	14	"	"	"	3/8x4x44	"	16x12
5/12 Slope	2x4	2x4	27	25	24	0	0	0	0	0	0	2x4	2x4	2x4	3/8x3½x11	8x12	8x8
	2x6	2x4	55	53	24	24	0	0	0	0	0	"	"	"	3/8x4x19	10x12	"
	2x6	2x6	54	52	50	23	21	20	0	0	0	"	"	"	3/8x4x20	"	"
	2x8	2x6	82	77	74	35	32	23	17	0	0	2x4	4&4	2x4	3/8x4x26	12x16	8x10
	2x10	4+4	100+	100+	100+	49	46	0	24	0	0	"	"	"	½x4x20	14x16	10x10
	2x12	4+6	-	-	-	63	58	52	31	22	0	"	"	"	½x4x22	18x16	12x12
	2x12	6+6	-	-	-	61	56	54	30	26	17	"	"	"	3/8x4x40	"	16x10

1400f Lumber

Slope	Top chord	Bottom chord	Truss spacing, ft. 2': Ceiling dead load, psf 0	2': 5	2': 8	4': 0	4': 5	4': 8	8': 0	8': 5	8': 8	Web member sizes W1	W2	W3	Gusset Sizes, in. A T H W	B H W	C H W
			---Max. snow + roof dead load, psf---														
3/12 Slope	2x4	2x4	28	25	24	12	0	0	0	0	0	2x4	2x4	2x4	3/8x3½x18	8x12	8x8
	2x6	2x4	55	52	34	24	0	0	0	0	0	"	"	"	½x4x16	10x16	8x10
	2x6	2x6	53	49	47	23	20	17	0	0	0	"	"	"	3/8x4x30	"	"
	2x8	2x6	76	68	65	33	28	25	0	0	0	2x4	2x4	2x4	½x4x22	12x16	8x12
	2x10	4+4	100+	100+	100+	47	43	22	23	0	0	"	4&4	"	½x4x30	14x16	12x10
	2x12	4+6	-	-	-	61	55	53	30	25	0	"	"	"	½x4x36	16x20	14x10
	2x12	6+6	-	-	-	58	53	50	29	24	21	"	"	"	3/8x4x58	"	16x12
4/12 Slope	2x4	2x4	31	29	28	13	0	0	0	0	0	2x4	2x4	2x4	3/8x3½x16	8x12	8x8
	2x6	2x4	63	59	49	27	0	0	0	0	0	"	"	"	½x4x15	10x12	8x10
	2x6	2x6	61	58	56	26	24	22	13	0	0	"	4&4	"	3/8x4x27	12x12	10x10
	2x8	2x6	91	85	82	40	36	34	20	13	0	2x4	4&4	2x4	½x4x20	12x16	10x10
	2x10	4+4	100+	100+	100+	54	50	32	27	0	0	"	"	"	½x4x27	16x16	12x12
	2x12	4+6	-	-	-	70	64	61	35	30	13	"	"	"	½x4x30	16x20	14x12
	2x12	6+6	-	-	-	67	62	59	33	29	27	"	"	"	½x4x34	"	18x12
5/12 Slope	2x4	2x4	33	32	31	14	0	0	0	0	0	2x4	2x4	2x4	3/8x3½x14	8x12	8x8
	2x6	2x4	69	66	53	30	0	0	0	0	0	"	"	"	½x4x13	10x12	"
	2x6	2x6	67	64	62	29	27	25	14	12	0	"	4&4	"	3/8x4x24	10x16	8x10
	2x8	2x6	100+	95	92	44	41	39	22	15	0	2x4	4&4	2x4	½x4x18	12x16	8x10
	2x10	4+4	-	100+	100+	61	56	44	30	0	0	"	"	"	½x4x25	16x16	10x12
	2x12	4+6	-	-	-	77	71	68	38	34	18	"	"	"	½x4x27	16x20	12x12
	2x12	6+6	-	-	-	75	69	66	37	33	31	"	"	"	½x4x28	"	16x12

1600f Lumber

Slope	Top chord	Bottom chord	Truss spacing, ft. 2': Ceiling dead load, psf 0	2': 5	2': 8	4': 0	4': 5	4': 8	8': 0	8': 5	8': 8	Web member sizes W1	W2	W3	Gusset Sizes, in. A T H W	B H W	C H W
			---Max. snow + roof dead load, psf---														
3/12 Slope	2x4	2x4	33	31	29	14	0	0	0	0	0	2x4	2x4	2x4	3/8x3½x21	8x12	8x8
	2x6	2x4	66	62	51	28	0	0	0	0	0	"	"	"	½x4x19	10x16	8x10
	2x6	2x6	63	59	57	27	24	23	13	0	0	"	"	"	3/8x4x36	"	8x12
	2x8	2x6	90	82	82	39	34	31	19	14	0	2x4	2x4	2x4	½x4x27	12x20	10x10
	2x10	4+4	101+	101+	101+	57	52	40	28	0	0	"	"	"	½x4x36	14x20	12x12
	2x12	4+6	-	-	-	74	67	63	37	32	17	"	"	"	½x4x40	16x20	14x12
	2x12	6+6	-	-	-	70	64	60	35	30	28	"	"	"	½x4x44	18x24	16x14
4/12 Slope	2x4	2x4	37	35	34	16	14	0	0	0	0	2x4	2x4	2x4	3/8x3½x18	8x12	8x10
	2x6	2x4	75	71	66	32	14	0	0	0	0	"	"	"	½x4x17	10x16	10x10
	2x6	2x6	73	69	67	32	29	27	16	13	0	"	"	"	½x4x18	"	"
	2x8	2x6	101+	101+	99	47	44	43	23	20	0	2x4	2x4	2x4	½x4x24	14x16	10x12
	2x10	4+4	-	-	-	66	61	50	33	14	0	"	"	"	½x4x28	14x20	12x12
	2x12	4+6	-	-	-	84	77	73	42	37	26	"	"	"	½x4x36	18x20	16x14
	2x12	6+6	-	-	-	81	74	71	40	36	34	"	"	"	½x4x36	"	18x14
5/12 Slope	2x4	2x4	39	38	38	17	15	0	0	0	0	2x4	2x4	2x4	3/8x3½x16	8x12	8x8
	2x6	2x4	87	78	78	35	25	0	17	0	0	"	"	"	½x4x16	10x16	8x10
	2x6	2x6	80	76	74	35	32	31	17	15	0	"	"	"	½x4x15	"	"
	2x8	2x6	101+	101+	101+	52	49	48	26	23	0	2x4	2x4	2x4	½x4x21	14x16	10x10
	2x10	4+4	-	-	-	73	68	59	36	19	0	"	"	"	½x4x25	14x20	10x12
	2x12	4+6				93	86	86	46	43	31	"	"	"	½x4x33	18x20	14x12
	2x12	6+6	-	-	-	90	83	80	45	41	38	"	"	"	½x4x33	"	18x12

32' SPAN, 2-WEB

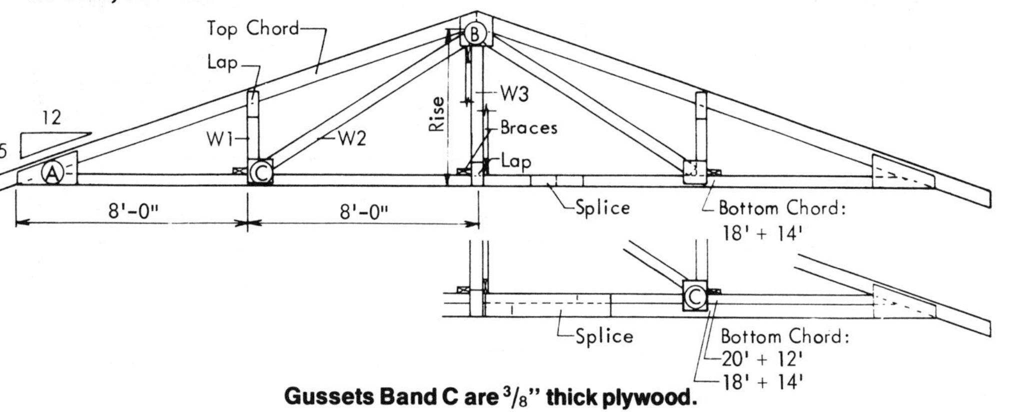

Gussets B and C are 3/8" thick plywood.

4+4, 4+6, 6+6 indicates stacked lower chord.
4&4, 6&4, indicate double web; a 2x4 is attached to the web member to increase its stiffness.

Before selecting heel gusset A, see pages 8 and 9.

Web Lengths

Roof Slope	Rise	Top Chord	W1	W2	W3
3/12	4'-0"	17'	2'	9'+8'	4'
4/12	5'-4"	18'	3'	10'+9'	5'
5/12	6'-8"	18'	3'	10'+9'	7'

1100f Lumber

			Truss spacing, ft.													
			2'			4'			8'			Web member sizes			Gusset Sizes, in.		
			Ceiling dead load, psf												A	B	C
	Top chord	Bottom chord	0	5	8	0	5	8	0	5	8	W1	W2	W3	T H W	H W	H W
			---Max. snow + roof dead load, psf---														
3/12 Slope	2x4	2x4	20	18	0	0	0	0	0	0	0	2x4	2x4	2x4	3/8x3½x14	8x12	8x8
	2x6	2x4	40	30	0	0	0	0	0	0	0	"	"	"	3/8x4x23	10x12	8x8
	2x6	2x6	39	36	34	17	14	0	0	0	0	"	"	"	3/8x4x25	10x16	8x10
	2x8	2x6	53	48	46	23	18	0	0	0	0	2x4	2x4	2x4	3/8x4x31	12x16	8x10
	2x10	4+4	80	74	68	35	20	0	0	0	0	"	"	"	½x4x28	14x16	10x10
	2x12	4+6	100+	98	93	45	39	30	22	0	0	"	4&4	"	½x4x36	16x16	12x12
	2x12	6+6	-	-	90	43	38	35	21	16	0	"	"	"	3/8x4x53	"	16x10
4/12 Slope	2x4	2x4	22	21	0	0	0	0	0	0	0	2x4	2x4	2x4	3/8x3½x12	8x12	8x8
	2x6	2x4	46	40	0	20	0	0	0	0	0	"	"	"	3/8x4x20	10x12	8x10
	2x6	2x6	45	42	41	19	17	0	0	0	0	"	"	"	3/8x4x22	"	"
	2x8	2x6	67	63	61	29	26	0	0	0	0	2x4	2x4	2x4	3/8x4x29	14x12	10x10
	2x10	4+4	93	86	82	40	30	0	20	0	0	"	4&4	"	½x4x22	14x16	12x10
	2x12	4+6	100+	100+	100+	52	47	38	26	13	0	"	"	"	½x4x28	18x16	14x12
	2x12	6+6	-	-	-	50	46	44	25	21	0	"	"	"	3/8x4x44	"	16x12
5/12 Slope	2x4	2x4	24	22	0	0	0	0	0	0	0	2x4	2x4	2x4	3/8x3½x10	8x12	8x8
	2x6	2x4	50	48	0	22	0	0	0	0	0	"	"	"	3/8x4x19	10x12	"
	2x6	2x6	49	47	46	21	19	15	0	0	0	"	"	"	3/8x4x19	"	"
	2x8	2x6	74	70	68	32	29	15	0	0	0	2x4	2x4	2x4	3/8x4x26	12x16	8x10
	2x10	4+4	100+	96	95	45	38	0	22	0	0	"	4&4	"	½x4x19	14x16	10x10
	2x12	4+6	-	100+	100+	57	53	46	28	16	0	"	"	"	½x4x24	18x16	14x10
	2x12	6+6	-	-	-	56	51	49	28	24	0	"	"	"	3/8x4x40	"	16x10

1400f Lumber

Slope	Top chord	Bottom chord	Truss spacing, ft. 2': Ceiling dead load, psf 0	2': 5	2': 8	4': 0	4': 5	4': 8	8': 0	8': 5	8': 8	Web member sizes W1	W2	W3	Gusset Sizes, in. A T H W	B H W	C H W
			---Max. snow + roof dead load, psf---														
3/12 Slope	2x4	2x4	25	23	21	0	0	0	0	0	0	2x4	2x4	2x4	3/8x3½x18	8x12	8x8
	2x6	2x4	50	47	22	0	0	0	0	0	0	"	"	"	½x4x16	10x16	8x10
	2x6	2x6	48	45	43	21	18	16	0	0	0	"	"	"	3/8x4x30	"	"
	2x8	2x6	70	63	60	30	25	19	0	0	0	2x4	2x4	2x4	½x4x22	12x16	8x12
	2x10	4+4	99	91	91	43	38	13	0	0	0	"	"	"	½x4x29	14x20	12x10
	2x12	4+6	-	100+	100+	55	50	48	27	20	0	"	4&4	"	½x4x36	16x20	14x12
	2x12	6+6	-	-	-	52	48	45	26	22	17	"	"	"	3/8x4x58	"	16x12
4/12 Slope	2x4	2x4	28	26	25	12	0	0	0	0	0	2x4	2x4	2x4	3/8x3½x15	8x12	8x8
	2x6	2x4	57	54	34	25	11	0	0	0	0	"	"	"	½x4x14	10x12	8x10
	2x6	2x6	56	53	51	24	21	20	0	0	0	"	"	"	3/8x4x26	12x12	10x10
	2x8	2x6	83	78	75	36	33	27	0	0	0	2x4	2x4	2x4	½x4x20	12x16	10x10
	2x10	4+4	100+	100+	100+	50	46	18	25	0	0	"	4&4	"	½x4x27	16x16	12x12
	2x12	4+6	-	-	-	64	58	56	32	26	0	"	"	"	½x4x29	16x20	14x12
	2x12	6+6	-	-	-	62	56	54	31	26	23	"	"	"	½x4x34	"	18x12
5/12 Slope	2x4	2x4	30	28	28	13	0	0	0	0	0	2x4	2x4	2x4	3/8x3½x13	8x12	8x8
	2x6	2x4	62	60	45	27	0	0	0	0	0	"	"	"	½x4x13	10x12	"
	2x6	2x6	61	58	57	26	24	23	13	0	0	"	4&4	"	3/8x4x23	10x16	8x10
	2x8	2x6	92	87	84	40	38	34	20	0	0	2x4	4&4	2x4	½x4x18	12x16	8x10
	2x10	4+4	100+	100+	100+	55	52	26	27	0	0	"	"	"	½x4x24	16x16	10x12
	2x12	4+6	-	-	-	71	65	61	35	31	0	"	"	"	½x4x27	16x20	12x12
	2x12	6+6	-	-	-	69	64	61	34	30	28	"	"	"	½x4x28	18x20	16x12

1600f Lumber

Slope	Top chord	Bottom chord	Truss spacing, ft. 2': Ceiling dead load, psf 0	2': 5	2': 8	4': 0	4': 5	4': 8	8': 0	8': 5	8': 8	Web member sizes W1	W2	W3	Gusset Sizes, in. A T H W	B H W	C H W
			---Max. snow + roof dead load, psf---														
3/12 Slope	2x4	2x4	30	28	26	13	0	0	0	0	0	2x4	2x4	2x4	3/8x3½x21	8x12	8x8
	2x6	2x4	60	57	42	26	0	0	0	0	0	"	"	"	½x4x19	10x16	8x10
	2x6	2x6	58	54	52	25	22	20	0	0	0	"	"	"	3/8x4x35	"	8x12
	2x8	2x6	84	76	72	36	31	29	0	0	0	2x4	2x4	2x4	½x4x26	12x16	10x10
	2x10	4+4	100+	100+	100+	52	48	26	26	0	0	"	4&4	"	½x4x35	14x20	12x12
	2x12	4+6	-	-	-	67	61	58	33	29	0	"	"	"	½x4x39	16x20	14x12
	2x12	6+6	-	-	-	63	58	55	32	27	24	"	"	"	½x4x44	18x24	16x14
4/12 Slope	2x4	2x4	33	32	30	14	0	0	0	0	0	2x4	2x4	2x4	3/8x3½x18	8x12	8x10
	2x6	2x4	68	65	51	29	17	0	0	0	0	"	"	"	½x4x17	10x16	10x10
	2x6	2x6	66	63	61	29	26	25	14	0	0	"	4&4	"	½x4x18	"	"
	2x8	2x6	100	94	90	43	41	38	21	15	0	2x4	4&4	2x4	½x4x23	14x16	10x12
	2x10	4+4	-	100+	100+	60	55	41	30	0	0	"	"	"	½x4x32	14x20	14x12
	2x12	4+6	-	-	-	76	70	66	38	34	15	"	"	"	½x4x35	18x20	16x14
	2x12	6+6	-	-	-	74	68	65	37	33	30	"	"	"	½x4x36	18x24	18x14
5/12 Slope	2x4	2x4	35	34	34	15	0	0	0	0	0	2x4	2x4	2x4	3/8x3½x16	8x12	8x8
	2x6	2x4	74	71	63	32	0	0	0	0	0	"	"	"	½x4x15	10x16	8x10
	2x6	2x6	73	69	68	31	29	28	15	13	0	"	4&4	"	½x4x15	"	"
	2x8	2x6	100+	100+	100	48	45	43	29	18	0	2x4	4&4	2x4	½x4x21	14x16	10x10
	2x10	4+4	-	-	-	66	62	49	33	0	0	"	"	"	½x4x25	14x20	10x12
	2x12	4+6	-	-		85	78	78	42	38	23	"	"	"	½x4x32	18x20	14x12
	2x12	6+6	-	-	-	83	76	73	41	37	35	"	"	"	½x4x33	"	18x12

34' SPAN, 2-WEB

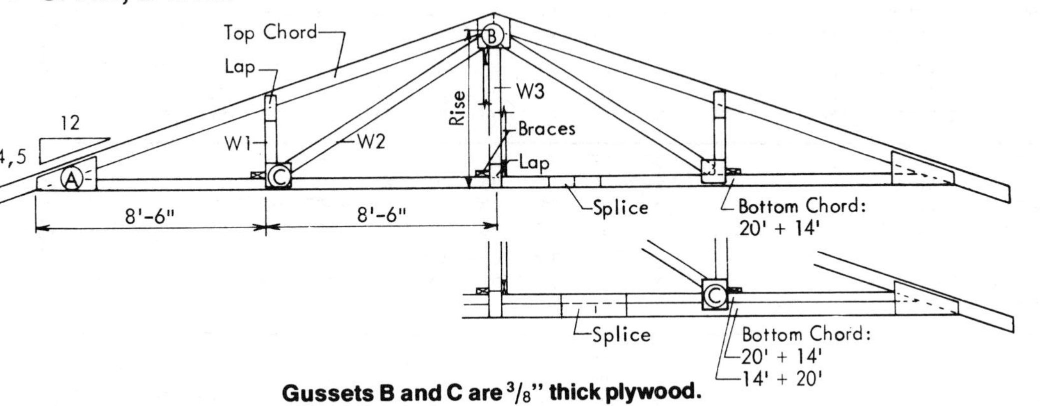

Gussets B and C are 3/8" thick plywood.

4+4, 4+6, 6+6 indicates stacked lower chord.
4&4, 6&4, indicate double web; a 2x4 is attached to the web member to increase its stiffness.

Before selecting heel gusset A, see pages 8 and 9 .

Web Lengths

Roof Slope	Rise	Top Chord	W1	W2	W3
3/12	4'-3"	18'	2'	9'+8'	4'
4/12	5'-8"	19'	3'	10'+9'	6'
5/12	7'-1"	19'	4'	11'+10'	7'

1100f Lumber

See Ceiling Load Note, page 4.

			Truss spacing, ft.									Web member sizes			Gusset Sizes, in.		
			2'			4'			8'								
			Ceiling dead load, psf												A	B	C
	Top chord	Bottom chord	0	5	8	0	5	8	0	5	8	W1	W2	W3	T H W	H W	H W
			---Max. snow + roof dead load, psf---														
3/12 Slope	2x4	2x4	18	16	0	0	0	0	0	0	0	2x4	2x4	2x4	3/8x3½x14	8x12	8x8
	2x6	2x4	37	22	0	0	0	0	0	0	0	"	"	"	3/8x4x22	10x12	"
	2x6	2x6	36	33	31	0	12	0	0	0	0	"	"	"	3/8x4x25	10x16	8x10
	2x8	2x6	49	45	42	0	16	0	0	0	0	2x4	2x4	2x4	3/8x4x31	12x16	8x10
	2x10	4+4	74	68	58	32	14	0	0	0	0	"	"	"	3/8x4x42	14x16	10x12
	2x12	4+6	94	86	86	41	36	23	0	0	0	"	"	"	½x4x36	16x16	14x10
	2x12	6+6	90	82	78	39	34	32	0	14	0	"	"	"	3/8x4x52	"	16x10
4/12 Slope	2x4	2x4	20	19	0	0	0	0	0	0	0	2x4	2x4	2x4	3/8x3½x12	8x12	8x8
	2x6	2x4	42	29	0	0	0	0	0	0	0	"	"	"	3/8x4x19	10x12	"
	2x6	2x6	41	39	37	18	15	0	0	0	0	"	"	"	3/8x4x22	"	8x10
	2x8	2x6	62	58	56	27	22	0	0	0	0	2x4	2x4	2x4	3/8x4x29	14x12	10x10
	2x10	4+4	85	79	76	37	20	0	0	0	0	"	"	"	½x4x22	14x16	12x10
	2x12	4+6	100+	100	100+	47	43	30	23	0	0	"	4&4	"	½x4x28	18x16	14x12
	2x12	6+6	–	–	–	46	42	39	23	19	0	"	"	"	3/8x4x44	"	16x12
5/12 Slope	2x4	2x4	22	20	0	0	0	0	0	0	0	2x4	2x4	2x4	3/8x3½x10	8x12	8x8
	2x6	2x4	46	40	0	0	0	0	0	0	0	"	"	"	3/8x4x17	10x12	"
	2x6	2x6	45	43	42	19	17	0	0	0	0	"	"	"	3/8x4x19	"	"
	2x8	2x6	68	64	62	29	27	0	0	0	0	2x4	2x4	2x4	3/8x4x26	12x16	8x10
	2x10	4+4	95	88	89	41	31	0	20	0	0	"	4&4	"	½x4x20	14x16	10x10
	2x12	4+6	100+	100+	100+	52	48	38	26	12	0	"	"	"	½x4x24	18x16	12x12
	2x12	6+6	–	–	–	51	47	45	25	22	14	"	"	"	3/8x4x40	"	14x12

1400f Lumber

Slope	Top chord	Bottom chord	Truss spacing, ft. 2' — Ceiling dead load, psf 0	2' 5	2' 8	4' 0	4' 5	4' 8	8' 0	8' 5	8' 8	Web member sizes W1	W2	W3	Gusset Sizes, in. A T H W	B H W	C H W
			---Max. snow + roof dead load, psf---														
3/12 Slope	2x4	2x4	23	21	13	0	0	0	0	0	0	2x4	2x4	2x4	3/8x3½x17	8x12	8x8
	2x6	2x4	46	43	13	0	0	0	0	0	0	"	"	"	½x4x16	10x16	8x10
	2x6	2x6	44	41	39	19	16	14	0	0	0	"	"	"	3/8x4x30	"	"
	2x8	2x6	65	59	56	28	23	14	0	0	0	2x4	2x4	2x4	½x4x22	12x16	8x12
	2x10	4+4	90	83	83	39	33	0	0	0	0	"	"	"	½x4x29	14x20	12x10
	2x12	4+6	100+	100+	100+	50	46	42	0	16	0	"	"	"	½x4x36	16x20	14x12
	2x12	6+6	-	-	-	48	44	41	24	20	0	"	4&4	"	3/8x4x58	"	16x12
4/12 Slope	2x4	2x4	25	24	20	0	0	0	0	0	0	2x4	2x4	2x4	3/8x3½x15	8x12	8x8
	2x6	2x4	52	50	20	0	0	0	0	0	0	"	"	"	½x4x14	10x12	8x10
	2x6	2x6	51	48	47	22	19	18	0	0	0	"	"	"	3/8x4x26	10x16	10x10
	2x8	2x6	76	72	69	33	30	22	0	0	0	2x4	2x4	2x4	½x4x20	12x16	10x10
	2x10	4+4	100+	98	98	46	42	0	0	0	0	"	"	"	½x4x26	16x16	12x12
	2x12	4+6	-	-	-	58	54	51	29	20	0	"	4&4	"	½x4x29	16x20	14x12
	2x12	6+6	-	-	-	57	52	50	28	24	16	"	"	"	½x4x34	"	18x12
5/12 Slope	2x4	2x4	27	26	25	12	0	0	0	0	0	2x4	2x4	2x4	3/8x3½x13	8x12	8x8
	2x6	2x4	57	54	26	24	0	0	0	0	0	"	"	"	3/8x4x22	10x12	"
	2x6	2x6	56	53	52	24	22	20	0	0	0	"	"	"	3/8x4x23	10x16	8x10
	2x8	2x6	84	80	77	36	34	26	0	0	0	2x4	2x4	2x4	½x4x18	12x16	8x10
	2x10	4+4	100+	100+	100+	51	47	12	25	0	0	"	4&4	"	½x4x24	16x16	10x12
	2x12	4+6	-	-	-	65	60	56	32	26	0	"	"	"	½x4x26	16x20	12x12
	2x12	6+6	-	-	-	63	59	56	31	28	21	"	"	"	½x4x28	"	16x12

1600f Lumber

Slope	Top chord	Bottom chord	Truss spacing, ft. 2' — Ceiling dead load, psf 0	2' 5	2' 8	4' 0	4' 5	4' 8	8' 0	8' 5	8' 8	Web member sizes W1	W2	W3	Gusset Sizes, in. A T H W	B H W	C H W
			---Max. snow + roof dead load, psf---														
3/12 Slope	2x4	2x4	27	25	24	0	0	0	0	0	0	2x4	2x4	2x4	3/8x3½x21	8x12	8x8
	2x6	2x4	55	52	28	0	0	0	0	0	0	"	"	"	½x4x19	10x16	8x10
	2x6	2x6	53	50	48	23	20	18	0	0	0	"	"	"	3/8x4x35	"	8x12
	2x8	2x6	77	70	67	33	29	24	0	0	0	2x4	2x4	2x4	½x4x26	12x20	10x10
	2x10	4+4	100+	100+	100+	47	44	18	0	0	0	"	"	"	½x4x35	14x20	12x12
	2x12	4+6	-	-	-	61	56	53	30	24	0	"	4&4	"	½x4x38	16x20	14x12
	2x12	6+6	-	-	-	58	53	50	29	24	20	"	"	"	½x4x44	18x24	16x14
4/12 Slope	2x4	2x4	30	28	27	13	0	0	0	0	0	2x4	2x4	2x4	3/8x3½x17	8x12	8x10
	2x6	2x4	62	59	43	27	0	0	0	0	0	"	"	"	½x4x17	10x16	10x10
	2x6	2x6	61	58	56	26	24	22	13	0	0	"	4&4	"	3/8x4x30	"	"
	2x8	2x6	91	86	83	40	37	33	20	0	0	2x4	4&4	2x4	½x4x23	14x16	10x12
	2x10	4+4	100+	100+	100+	55	51	23	27	0	0	"	"	"	½x4x31	14x20	14x12
	2x12	4+6	-	-	-	70	65	60	35	30	0	"	"	"	½x4x34	18x20	16x14
	2x12	6+6	-	-	-	68	63	60	34	30	27	"	"	"	½x4x36	18x24	18x14
5/12 Slope	2x4	2x4	32	31	30	14	0	0	0	0	0	2x4	2x4	2x4	3/8x3½x15	8x12	8x8
	2x6	2x4	67	65	50	29	0	0	0	0	0	"	"	"	½x4x15	10x16	8x10
	2x6	2x6	66	63	62	29	26	25	14	12	0	"	4&4	"	½x4x15	"	"
	2x8	2x6	100+	95	92	44	41	39	22	12	0	2x4	4&4	2x4	½x4x21	14x16	10x10
	2x10	4+4	-	100+	100+	61	57	39	30	0	0	"	"	"	½x4x28	14x20	12x12
	2x12	4+6	-	-	-	78	72	71	39	35	0	"	"	"	½x4x32	18x20	14x12
	2x12	6+6	-	-	-	76	70	67	38	34	32	"	"	"	½x4x33	18x24	18x12

36' SPAN, 2-WEB

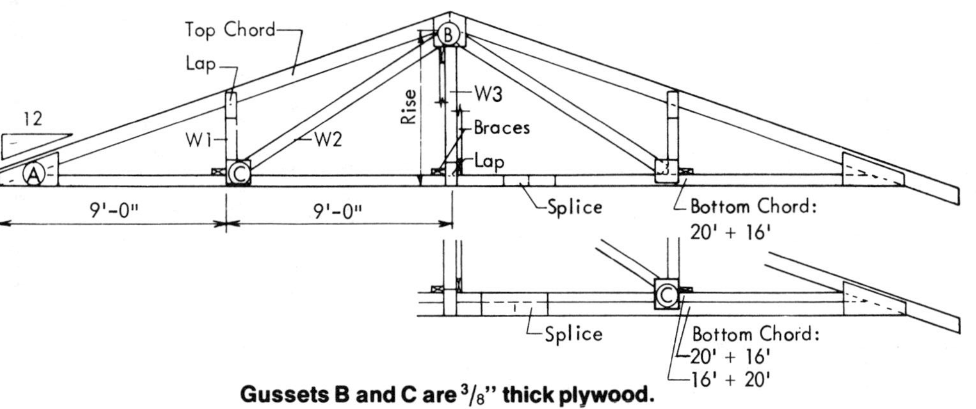

Gussets B and C are 3/8" thick plywood.

4+4, 4+6, 6+6 indicates stacked lower chord.
4&4, 6&4, indicate double web; a 2x4 is attached to the web member to increase its stiffness.

Before selecting heel gusset A, see pages 8 and 9 .

Web Lengths

Roof Slope	Rise	Top Chord	W1	W2	W3
3/12	4'-6"	19'	2'	10'+9'	5'
4/12	6'-0"	20'	3'	11'+10'	6'
5/12	7'-6"	20'	4'	12'+11'	8'

1100f Lumber

See Ceiling Load Note, page 4.

			Truss spacing, ft. 2'			4'			8'			Web member sizes			Gusset Sizes, in. A	B	C
			Ceiling dead load, psf														
Slope	Top chord	Bottom chord	0	5	8	0	5	8	0	5	8	W1	W2	W3	T H W	H W	H W
			---Max. snow + roof dead load, psf---														
3/12 Slope	2x4	2x4	17	15	0	0	0	0	0	0	0	2x4	2x4	2x4	3/8x3½x14	8x12	8x8
	2x6	2x4	34	15	0	0	0	0	0	0	0	"	"	"	3/8x4x22	10x12	"
	2x6	2x6	33	30	28	0	0	0	0	0	0	"	"	"	3/8x4x25	10x16	8x10
	2x8	2x6	46	42	38	0	12	0	0	0	0	2x4	2x4	2x4	3/8x4x31	12x16	8x10
	2x10	4+4	68	63	50	29	0	0	0	0	0	"	4&4	"	3/8x4x42	14x16	10x12
	2x12	4+6	87	79	79	38	31	15	0	0	0	"	"	"	½x4x36	16x16	14x10
	2x12	6+6	83	76	72	36	32	29	0	0	0	"	"	"	3/8x4x51	"	16x10
4/12 Slope	2x4	2x4	18	17	0	0	0	0	0	0	0	2x4	2x4	2x4	3/8x3½x12	8x12	8x8
	2x6	2x4	38	21	0	0	0	0	0	0	0	"	"	"	3/8x4x19	10x12	"
	2x6	2x6	38	36	34	16	14	0	0	0	0	"	4&4	"	3/8x4x21	"	8x10
	2x8	2x6	57	53	52	24	18	0	0	0	0	2x4	4&4	2x4	3/8x4x29	12x16	10x10
	2x10	4+4	79	73	65	34	12	0	0	0	0	"	"	"	½x4x22	14x16	12x10
	2x12	4+6	100+	93	93	44	40	22	0	0	0	"	"	"	½x4x28	18x16	14x12
	2x12	6+6	–	90	91	42	38	36	21	14	0	"	"	"	½x4x34	"	16x12
5/12 Slope	2x4	2x4	20	18	0	0	0	0	0	0	0	2x4	2x4	2x4	3/8x3½x10	8x12	8x8
	2x6	2x4	42	28	0	0	0	0	0	0	0	"	"	"	3/8x4x17	10x12	"
	2x6	2x6	41	39	38	18	16	0	0	0	0	"	4&4	"	3/8x4x19	"	"
	2x8	2x6	62	59	57	27	22	0	0	0	0	2x4	4&4	2x4	3/8x4x26	12x16	8x10
	2x10	4+4	87	81	78	38	16	0	0	0	0	"	"	"	½x4x19	14x16	10x10
	2x12	4+6	100+	100+	100+	48	45	31	24	0	0	"	"	"	½x4x24	18x16	12x12
	2x12	6+6	–	–	–	47	44	42	23	19	0	"	"	"	3/8x4x39	"	14x12

1400f Lumber

Slope	Top chord	Bottom chord	2' (0)	2' (5)	2' (8)	4' (0)	4' (5)	4' (8)	8' (0)	8' (5)	8' (8)	W1	W2	W3	A T H W	B H W	C H W
			Truss spacing, ft.									Web member sizes			Gusset Sizes, in.		
			Ceiling dead load, psf														
			---Max. snow + roof dead load, psf---														
3/12 Slope	2x4	2x4	21	19	0	0	0	0	0	0	0	2x4	2x4	2x4	3/8x3½x17	8x12	8x8
	2x6	2x4	42	36	0	0	0	0	0	0	0	"	"	"	3/8x4x27	10x16	8x10
	2x6	2x6	41	38	36	17	15	0	0	0	0	"	"	"	3/8x4x29	"	"
	2x8	2x6	62	55	52	26	21	0	0	0	0	2x4	2x4	2x4	½x4x22	12x16	8x12
	2x10	4+4	83	77	76	36	24	14	0	0	0	"	"	"	½x4x29	14x20	12x12
	2x12	4+6	100+	98	99	46	42	36	0	0	0	"	"	"	½x4x36	16x20	14x12
	2x12	6+6	-	-	-	44	40	37	22	17	0	"	4&4	"	3/8x4x57	"	16x12
4/12 Slope	2x4	2x4	23	21	0	0	0	0	0	0	0	2x4	2x4	2x4	3/8x3½x14	8x12	8x8
	2x6	2x4	48	45	0	0	0	0	0	0	0	"	"	"	½x4x14	10x12	8x10
	2x6	2x6	47	44	43	20	18	12	0	0	0	"	"	"	3/8x4x26	12x12	10x10
	2x8	2x6	70	66	64	30	27	13	0	0	0	2x4	2x4	2x4	½x4x19	12x16	10x10
	2x10	4+4	97	91	90	42	36	13	0	0	0	"	"	"	½x4x26	16x16	12x12
	2x12	4+6	100+	100+	100+	54	50	45	27	16	0	"	4&4	"	½x4x32	16x20	16x12
	2x12	6+6	-	-	-	52	48	46	26	22	12	"	"	"	½x4x34	"	18x12
5/12 Slope	2x4	2x4	25	23	12	0	0	0	0	0	0	2x4	2x4	2x4	3/8x3½x12	8x12	8x8
	2x6	2x4	52	50	12	22	0	0	0	0	0	"	4&4	"	3/8x4x22	10x12	"
	2x6	2x6	51	49	48	22	20	18	0	0	0	"	"	"	3/8x4x23	10x16	8x10
	2x8	2x6	78	73	71	34	31	18	0	0	0	2x4	4&4	2x4	½x4x17	12x16	8x10
	2x10	4+4	100+	100+	100+	47	44	0	23	0	0	"	"	"	½x4x24	16x16	10x12
	2x12	4+6	-	-	-	60	55	51	30	19	0	"	"	"	½x4x26	16x20	12x12
	2x12	6+6	-	-	-	59	54	52	29	25	19	"	"	"	½x4x28	"	16x12

1600f Lumber

Slope	Top chord	Bottom chord	2' (0)	2' (5)	2' (8)	4' (0)	4' (5)	4' (8)	8' (0)	8' (5)	8' (8)	W1	W2	W3	A T H W	B H W	C H W
			Truss spacing, ft.									Web member sizes			Gusset Sizes, in.		
			Ceiling dead load, psf														
			---Max. snow + roof dead load, psf---														
3/12 Slope	2x4	2x4	25	23	17	0	0	0	0	0	0	2x4	2x4	2x4	3/8x3½x20	8x12	8x8
	2x6	2x4	50	48	17	0	0	0	0	0	0	"	"	"	½x4x19	10x16	8x10
	2x6	2x6	49	46	44	21	18	16	0	0	0	"	"	"	3/8x4x34	"	8x12
	2x8	2x6	72	65	62	31	26	18	0	0	0	2x4	2x4	2x4	½x4x26	12x20	10x10
	2x10	4+4	100+	93	93	44	39	0	0	0	0	"	"	"	½x4x34	14x20	12x12
	2x12	4+6	-	100+	100+	56	51	49	28	19	0	"	"	"	½x4x36	18x24	16x14
	2x12	6+6		-	-	54	49	47	27	22	15	"	"	"	½x4x44	"	16x14
4/12 Slope	2x4	2x4	27	26	25	12	0	0	0	0	0	2x4	2x4	2x4	3/8x3½x17	8x12	8x10
	2x6	2x4	57	54	26	24	0	0	0	0	0	"	"	"	½x4x16	10x16	10x10
	2x6	2x6	56	53	52	24	22	20	0	0	0	"	"	"	3/8x4x30	"	"
	2x8	2x6	84	79	77	36	33	26	0	0	0	2x4	2x4	2x4	½x4x23	14x16	10x12
	2x10	4+4	100+	100+	100+	51	47	0	25	0	0	"	4&4	"	½x4x31	14x20	14x12
	2x12	4+6	-	-	-	65	60	55	32	25	0	"	"	"	½x4x33	18x20	16x14
	2x12	6+6	-	-	-	63	58	55	31	27	20	"	"	"	½x4x35	18x24	18x14
5/12 Slope	2x4	2x4	29	28	27	12	0	0	0	0	0	2x4	2x4	2x4	3/8x3½x15	8x12	8x8
	2x6	2x4	62	59	33	27	0	0	0	0	0	"	"	"	½x4x14	10x16	8x10
	2x6	2x6	61	58	57	26	24	23	13	0	0	"	4&4	"	3/8x4x26	"	"
	2x8	2x6	92	88	85	40	38	33	20	0	0	2x4	4&4	2x4	½x4x20	14x16	10x10
	2x10	4+4	100+	100+	100+	56	53	16	28	0	0	"	"	"	½x4x28	16x20	12x12
	2x12	4+6	-	-	-	72	66	65	36	30	0	"	"	"	½x4x31	18x20	14x12
	2x12	6+6	-	-	-	70	65	62	35	31	26	"	"	"	½x4x32	18x24	18x12

38' SPAN, 2-WEB

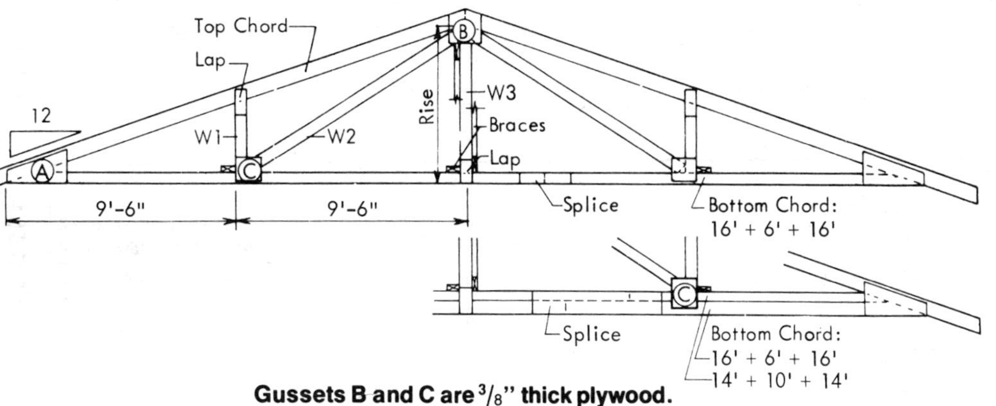

Gussets B and C are $^{3}/_{8}$" thick plywood.

4+4, 4+6, 6+6 indicates stacked lower chord.
4&4, 6&4, indicate double web; a 2x4 is attached to the web member to increase its stiffness.

Before selecting heel gusset A, see pages 8 and 9.

Web Lengths

Roof Slope	Rise	Top Chord	W1	W2	W3
3/12	4'-9"	20'	3'	11'+10'	5'
4/12	6'-4"	16'+5'	3'	11'+10'	6'
5/12	7'-11"	16'+5'	4'	12'+11'	8'

1100f Lumber

See Ceiling Load Note, page 4.

			Truss spacing, ft. 2'			4'			8'			Web member sizes			Gusset Sizes, in. A	B	C
			Ceiling dead load, psf														
	Top chord	Bottom chord	0	5	8	0	5	8	0	5	8	W1	W2	W3	T H W	H W	H W
			---Max. snow + roof dead load, psf---														
3/12 Slope	2x4	2x4	0	0	0	0	0	0	0	0	0	2x4	2x4	2x4	3/8x3½x14	8x12	8x8
	2x6	2x4	0	0	0	0	0	0	0	0	0	"	"	"	3/8x4x21	10x12	"
	2x6	2x6	30	27	26	0	0	0	0	0	0	"	"	"	3/8x4x24	10x16	8x10
	2x8	2x6	43	38	35	0	0	0	0	0	0	2x4	2x4	2x4	3/8x4x30	12x16	8x10
	2x10	4+4	63	58	43	0	0	0	0	0	0	"	"	"	3/8x4x41	14x16	10x12
	2x12	4+6	80	74	72	35	28	0	0	0	0	"	4&4	"	½x4x36	16x16	14x10
	2x12	6+6	77	71	67	33	29	26	0	0	0	"	"	"	3/8x4x51	"	16x10
4/12 Slope	2x4	2x4	17	15	0	0	0	0	0	0	0	2x4	2x4	2x4	3/8x3½x12	8x12	8x8
	2x6	2x4	35	15	0	0	0	0	0	0	0	"	"	"	3/8x4x19	10x12	"
	2x6	2x6	35	32	31	15	13	0	0	0	0	"	4&4	"	3/8x4x21	"	8x10
	2x8	2x6	52	49	48	23	14	0	0	0	0	2x4	4&4	2x4	3/8x4x29	12x16	10x10
	2x10	4+4	73	67	52	32	0	0	0	0	0	"	"	"	½x4x22	14x16	12x10
	2x12	4+6	93	86	86	40	35	12	0	0	0	"	"	"	½x4x28	18x16	14x12
	2x12	6+6	91	84	80	39	35	33	19	0	0	"	"	"	3/8x4x46	"	16x12
5/12 Slope	2x4	2x4	18	17	0	0	0	0	0	0	0	2x4	2x4	2x4	3/8x3½x10	8x12	8x8
	2x6	2x4	38	19	0	0	0	0	0	0	0	"	"	"	3/8x4x16	10x12	"
	2x6	2x6	38	36	34	16	14	0	0	0	0	"	4&4	"	3/8x4x19	"	"
	2x8	2x6	58	55	53	25	17	0	0	0	0	2x4	4&4	2x4	3/8x4x26	12x16	8x10
	2x10	4+4	81	75	65	35	21	0	0	0	0	"	"	"	½x4x19	14x16	10x10
	2x12	4+6	100+	96	97	45	42	21	0	0	0	"	"	"	½x4x24	18x16	12x12
	2x12	6+6	-	95	91	44	41	38	22	14	0	"	"	"	½x4x28	"	14x12

1400f Lumber

Slope	Top chord	Bottom chord	Truss spacing, ft. 2' — Ceiling dead load, psf 0	2' 5	2' 8	4' 0	4' 5	4' 8	8' 0	8' 5	8' 8	Web member sizes W1	W2	W3	Gusset Sizes, in. A T H W	B H W	C H W
			---Max. snow + roof dead load, psf---														
3/12Slope	2x4	2x4	19	17	0	0	0	0	0	0	0	2x4	2x4	2x4	3/8x3½x16	8x12	8x8
	2x6	2x4	39	27	0	0	0	0	0	0	0	"	"	"	3/8x4x26	10x12	"
	2x6	2x6	37	35	33	0	13	0	0	0	0	"	"	"	3/8x4x29	10x16	8x10
	2x8	2x6	55	51	49	0	19	0	0	0	0	2x4	2x4	2x4	½x4x22	12x16	8x12
	2x10	4+4	77	72	70	33	17	0	0	0	0	"	4&4	"	½x4x29	14x20	12x12
	2x12	4+6	99	90	91	43	38	28	0	0	0	"	"	"	½x4x36	16x20	14x12
	2x12	6+6	95	87	87	41	37	34	0	16	0	"	"	"	½x4x44	"	16x12
4/12 Slope	2x4	2x4	21	19	0	0	0	0	0	0	0	2x4	2x4	2x4	3/8x3½x14	8x12	8x8
	2x6	2x4	44	39	0	0	0	0	0	0	0	"	"	"	3/8x4x23	10x12	8x10
	2x6	2x6	43	41	40	18	16	0	0	0	0	"	4&4	"	3/8x4x26	12x12	10x10
	2x8	2x6	65	61	59	28	25	0	0	0	0	2x4	4&4	2x4	½x4x19	12x16	10x10
	2x10	4+4	90	84	81	39	26	0	0	0	0	"	"	"	½x4x25	16x16	12x12
	2x12	4+6	100+	100+	100+	50	46	38	25	0	0	"	"	"	½x4x32	16x20	16x12
	2x12	6+6	-	-	-	49	45	43	24	20	0	"	"	"	½x4x34	"	18x12
5/12 Slope	2x4	2x4	22	21	0	0	0	0	0	0	0	2x4	2x4	2x4	3/8x3½x12	8x12	8x8
	2x6	2x4	48	46	0	0	0	0	0	0	0	"	"	"	3/8x4x22	10x12	"
	2x6	2x6	47	45	44	20	18	0	0	0	0	"	4&4	"	3/8x4x23	10x16	8x10
	2x8	2x6	72	68	66	31	28	0	0	0	0	2x4	4&4	2x4	½x4x17	12x16	8x10
	2x10	4+4	100	94	94	43	32	0	0	0	0	"	"	"	½x4x23	16x16	10x12
	2x12	4+6	-	100+	100+	56	52	45	28	15	0	"	"	"	½x4x29	16x20	14x12
	2x12	6+6	-	-	-	55	51	48	27	23	16	"	"	"	½x4x28	"	16x12

1600f Lumber

Slope	Top chord	Bottom chord	Truss spacing, ft. 2' — Ceiling dead load, psf 0	2' 5	2' 8	4' 0	4' 5	4' 8	8' 0	8' 5	8' 8	Web member sizes W1	W2	W3	Gusset Sizes, in. A T H W	B H W	C H W
			---Max. snow + roof dead load, psf---														
3/12 Slope	2x4	2x4	23	21	0	0	0	0	0	0	0	2x4	2x4	2x4	3/8x3½x19	8x12	8x8
	2x6	2x4	46	43	0	0	0	0	0	0	0	"	"	"	½x4x18	10x16	8x10
	2x6	2x6	45	43	41	19	17	0	0	0	0	"	"	"	3/8x4x35	"	8x12
	2x8	2x6	67	61	58	29	24	0	0	0	0	2x4	2x4	2x4	½x4x26	12x20	10x12
	2x10	4+4	93	86	86	40	30	0	0	0	0	"	"	"	½x4x34	14x20	12x12
	2x12	4+6	100+	100+	100+	52	47	42	0	14	0	"	"	"	½x4x42	18x24	14x14
	2x12	6+6	-	-	-	50	45	43	25	20	0	"	4&4	"	½x4x44	"	16x14
4/12 Slope	2x4	2x4	25	23	12	0	0	0	0	0	0	2x4	2x4	2x4	3/8x3½x16	8x12	8x8
	2x6	2x4	52	50	12	22	0	0	0	0	0	"	"	"	½x4x16	10x16	10x10
	2x6	2x6	51	49	48	22	20	17	0	0	0	"	"	"	3/8x4x30	"	"
	2x8	2x6	78	73	71	34	31	17	0	0	0	2x4	2x4	2x4	½x4x23	14x16	10x12
	2x10	4+4	100+	100+	100+	47	43	0	23	0	0	"	4&4	"	½x4x30	16x20	14x14
	2x12	4+6	-	-	-	60	55	51	30	20	0	"	"	"	½x4x33	18x20	16x14
	2x12	6+6	-	-	-	58	54	52	29	25	15	"	"	"	½x4x35	"	18x14
5/12 Slope	2x4	2x4	26	25	17	0	0	0	0	0	0	2x4	2x4	2x4	3/8x3½x14	8x12	8x8
	2x6	2x4	57	55	17	24	0	0	0	0	0	"	"	"	½x4x14	10x16	8x10
	2x6	2x6	56	53	52	24	22	21	0	0	0	"	"	"	3/8x4x26	"	"
	2x8	2x6	85	81	79	37	34	25	0	0	0	2x4	2x4	2x4	½x4x20	14x16	10x10
	2x10	4+4	100+	100+	100+	52	49	0	26	0	0	"	4&4	"	½x4x28	16x20	12x12
	2x12	4+6	-	-	-	66	62	59	33	24	0	"	"	"	½x4x30	18x20	14x12
	2x12	6+6	-	-	-	65	61	58	32	29	19	"	"	"	½x4x31	"	18x12

40' SPAN, 2-WEB

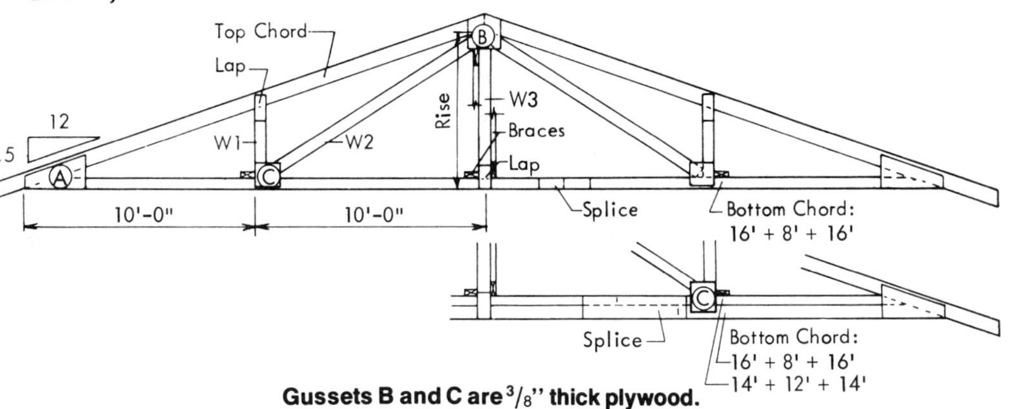

Gussets B and C are 3/8" thick plywood.

4+4, 4+6, 6+6 indicates stacked lower chord.
4&4, 6&4, indicate double web; a 2x4 is attached to the web member to increase its stiffness.

Before selecting heel gusset A, see pages 8 and 9 .

Web Lengths

Roof Slope	Rise	Top Chord	W1	W2	W3
3/12	5'-0"	16'+5'	3'	11'+10'	5'
4/12	6'-8"	16'+6'	3'	12'+11'	7'
5/12	8'-4"	16'+6'	4'	13'+12'	8'

1100f Lumber

See Ceiling Load Note, page 4.

Truss spacing, ft. (2', 4', 8'); Ceiling dead load, psf (0, 5, 8); ---Max. snow + roof dead load, psf---

Slope	Top chord	Bottom chord	2' 0	2' 5	2' 8	4' 0	4' 5	4' 8	8' 0	8' 5	8' 8	Web W1	Web W2	Web W3	Gusset A T H W	Gusset B H W	Gusset C H W
3/12 Slope	2x4	2x4	0	0	0	0	0	0	0	0	0	2x4	2x4	2x4	3/8x3½x14	8x12	8x8
	2x6	2x4	0	0	0	0	0	0	0	0	0	"	"	"	3/8x4x21	10x12	"
	2x6	2x6	28	25	24	0	0	0	0	0	0	"	"	"	3/8x4x24	10x16	8x10
	2x8	2x6	40	36	31	0	0	0	0	0	0	2x4	2x4	2x4	3/8x4x30	12x16	8x10
	2x10	4+4	58	54	30	0	0	0	0	0	0	"	"	"	3/8x4x41	14x16	10x12
	2x12	4+6	75	68	66	32	24	15	0	0	0	"	4&4	"	3/8x4x50	16x16	14x10
	2x12	6+6	72	66	63	31	27	20	0	0	0	"	"	"	3/8x4x50	"	16x10
4/12 Slope	2x4	2x4	15	0	0	0	0	0	0	0	0	2x4	2x4	2x4	3/8x3½x12	8x12	8x8
	2x6	2x4	33	0	0	0	0	0	0	0	0	"	"	"	3/8x4x18	10x12	"
	2x6	2x6	32	30	28	0	0	0	0	0	0	"	"	"	3/8x4x21	"	8x10
	2x8	2x6	49	46	43	0	0	0	0	0	0	2x4	2x4	2x4	3/8x4x28	14x12	10x10
	2x10	4+4	68	62	45	29	0	0	0	0	0	"	4&4	"	½x4x22	14x16	12x10
	2x12	4+6	87	80	79	38	30	0	0	0	0	"	"	"	½x4x28	18x16	14x12
	2x12	6+6	85	78	75	37	33	28	0	0	0	"	"	"	3/8x4x45	"	16x12
5/12 Slope	2x4	2x4	16	0	0	0	0	0	0	0	0	2x4	4&4	2x4	3/8x3½x10	8x12	8x8
	2x6	2x4	35	0	0	0	0	0	0	0	0	"	"	"	3/8x4x16	10x12	"
	2x6	2x6	35	33	32	0	12	0	0	0	0	"	"	"	3/8x4x18	"	"
	2x8	2x6	53	51	49	0	13	0	0	0	0	2x4	4&4	2x4	3/8x4x25	12x16	8x10
	2x10	4+4	75	69	51	33	0	0	0	0	0	"	"	"	½x4x18	14x16	10x10
	2x12	4+6	96	90	90	42	36	0	0	0	0	"	"	"	½x4x23	18x16	12x12
	2x12	6+6	95	88	85	41	37	34	0	0	0	"	"	"	½x4x28	"	14x12

1400f Lumber

Slope	Top chord	Bottom chord	2' / 0	2' / 5	2' / 8	4' / 0	4' / 5	4' / 8	8' / 0	8' / 5	8' / 8	W1	W2	W3	A T H W	B H W	C H W
			Truss spacing, ft. — Ceiling dead load, psf — ---Max. snow + roof dead load, psf---									Web member sizes			Gusset Sizes, in.		
3/12 Slope	2x4	2x4	0	15	0	0	0	0	0	0	0	2x4	2x4	2x4	3/8x3½x16	8x12	8x8
	2x6	2x4	0	21	0	0	0	0	0	0	0	"	"	"	3/8x4x26	10x12	"
	2x6	2x6	35	32	30	0	12	0	0	0	0	"	"	"	3/8x4x29	10x16	8x10
	2x8	2x6	52	48	46	0	15	0	0	0	0	2x4	2x4	2x4	½x4x22	12x16	10x10
	2x10	4+4	72	67	59	0	0	0	0	0	0	"	"	"	½x4x28	14x20	12x10
	2x12	4+6	92	84	84	40	35	22	0	0	0	"	4&4	"	½x4x36	16x20	14x12
	2x12	6+6	88	81	77	38	34	31	0	14	0	"	"	"	½x4x44	"	16x12
4/12 Slope	2x4	2x4	19	18	0	0	0	0	0	0	0	2x4	2x4	2x4	3/8x3½x14	8x12	8x8
	2x6	2x4	41	28	15	0	0	0	0	0	0	"	"	"	3/8x4x22	10x12	8x10
	2x6	2x6	40	38	36	0	15	0	0	0	0	"	"	"	3/8x4x25	12x12	10x10
	2x8	2x6	60	57	55	0	22	0	0	0	0	2x4	2x4	2x4	½x4x19	12x16	10x10
	2x10	4+4	84	78	77	36	17	0	0	0	0	"	4&4	"	½x4x25	14x20	12x12
	2x12	4+6	100+	99	100	46	43	30	0	0	0	"	"	"	½x4x31	16x20	16x12
	2x12	6+6	-	97	98	45	42	39	0	19	0	"	"	"	½x4x34	18x20	18x14
5/12 Slope	2x4	2x4	20	19	0	0	0	0	0	0	0	2x4	4&4	2x4	3/8x3½x12	8x12	8x8
	2x6	2x4	44	39	0	0	0	0	0	0	0	"	"	"	3/8x4x20	10x12	"
	2x6	2x6	43	41	41	19	17	0	0	0	0	"	"	"	3/8x4x22	10x16	8x10
	2x8	2x6	66	63	61	29	26	0	0	0	0	2x4	4&4	2x4	½x4x17	12x16	8x10
	2x10	4+4	93	88	88	40	23	0	0	0	0	"	"	"	½x4x23	16x16	10x12
	2x12	4+6	100+	100+	100+	52	48	38	0	0	0	"	"	"	½x4x29	18x20	14x12
	2x12	6+6	-	-	-	51	47	45	25	22	13	"	"	"	½x4x28	"	16x12

1600f Lumber

Slope	Top chord	Bottom chord	2' / 0	2' / 5	2' / 8	4' / 0	4' / 5	4' / 8	8' / 0	8' / 5	8' / 8	W1	W2	W3	A T H W	B H W	C H W
			Truss spacing, ft. — Ceiling dead load, psf — ---Max. snow + roof dead load, psf---									Web member sizes			Gusset Sizes, in.		
3/12 Slope	2x4	2x4	21	19	0	0	0	0	0	0	0	2x4	2x4	2x4	3/8x3½x19	8x12	8x8
	2x6	2x4	43	34	0	0	0	0	0	0	0	"	"	"	½x4x17	10x16	8x10
	2x6	2x6	42	39	37	0	15	0	0	0	0	"	"	"	3/8x4x34	"	8x12
	2x8	2x6	62	57	54	0	22	0	0	0	0	2x4	2x4	2x4	½x4x26	12x20	10x12
	2x10	4+4	87	80	80	37	22	0	0	0	0	"	"	"	½x4x34	16x24	12x12
	2x12	4+6	100+	102	103	48	44	36	0	0	0	"	"	"	½x4x42	18x24	14x14
	2x12	6+6	-	98	99	46	42	39	0	19	0	"	"	"	½x4x44	"	16x14
4/12 Slope	2x4	2x4	23	21	0	0	0	0	0	0	0	2x4	2x4	2x4	3/8x3½x16	8x12	8x8
	2x6	2x4	48	46	0	0	0	0	0	0	0	"	"	"	½x4x16	10x16	10x10
	2x6	2x6	47	45	44	20	18	0	0	0	0	"	"	"	3/8x4x30	"	"
	2x8	2x6	72	68	66	31	28	0	0	0	0	2x4	2x4	2x4	½x4x22	14x16	10x12
	2x10	4+4	100+	94	92	44	33	0	0	0	0	"	"	"	½x4x30	14x20	14x12
	2x12	4+6	-	100+	100+	56	52	45	28	14	0	"	4&4	"	½x4x34	18x20	16x14
	2x12	6+6	-	-	-	54	50	48	27	23	0	"	"	"	½x4x34	18x24	18x14
5/12 Slope	2x4	2x4	24	23	0	0	0	0	0	0	0	2x4	4&4	2x4	3/8x3½x13	8x12	8x8
	2x6	2x4	52	50	0	0	0	0	0	0	0	"	"	"	½x4x14	10x16	8x10
	2x6	2x6	51	49	48	22	20	12	0	0	0	"	"	"	3/8x4x26	"	"
	2x8	2x6	79	75	73	34	32	12	0	0	0	2x4	4&4	2x4	½x4x20	14x16	10x10
	2x10	4+4	100+	100+	100+	48	43	0	0	0	0	"	"	"	½x4x27	16x20	12x12
	2x12	4+6	-	-	-	62	58	51	31	19	0	"	"	"	½x4x30	18x20	14x12
	2x12	6+6	-	-	-	61	57	54	31	27	13	"	"	"	½x4x31	"	18x12

36' SPAN, 4-WEB

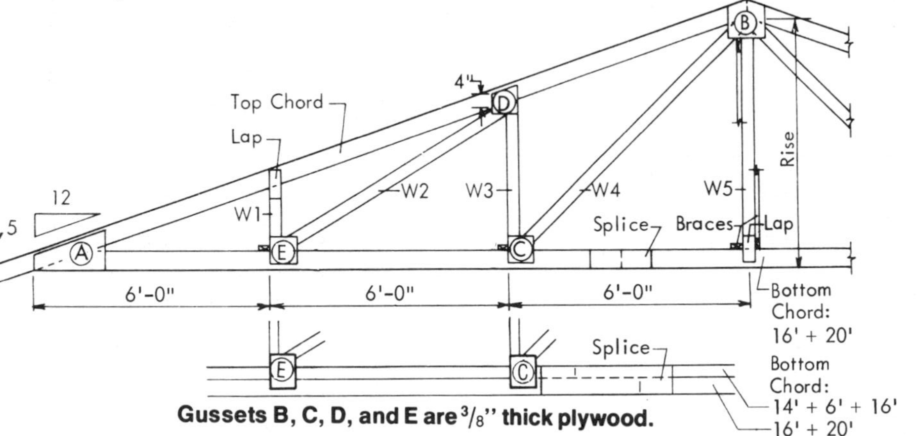

Gussets B, C, D, and E are 3/8" thick plywood.

4+4, 4+6, 6+6 indicates stacked lower chord.
4&4, 6&4, indicate double web; a 2x4 is attached to the web member to increase its stiffness.

Before selecting heel gusset A, see pages 8 and 9 .

Web Lengths

Roof Slope	Rise	Top Chord	W1	W2	W3	W4	W5
3/12	4'-6"	19'	2'	7'	3'	7'	5'
4/12	6'-0"	20'	2'	7'	4'	8'	6'
5/12	7'-6"	20'	3'	8'	5'	10'	8'

1100f Lumber

Truss spacing, ft.: 2', 4', 8'; Ceiling dead load, psf: 0, 5, 8; ---Max. snow + roof dead load, psf---

Slope	Top chord	Bottom chord	2' / 0	2' / 5	2' / 8	4' / 0	4' / 5	4' / 8	8' / 0	8' / 5	8' / 8	W1	W2	W3	W4	W5	A T H W	B H W	C H W	D H W	E H W
3/12 Slope	2x4	2x4	26	23	21	0	0	0	0	0	0	2x4	2x4	2x4	2x4	2x4	3/8x3½x21	8x12	8x8	8x8	8x8
	2x6	2x4	31	28	25	0	0	0	0	0	0	"	"	"	"	"	3/8x4x22	10x12	"	"	"
	2x6	2x6	45	41	38	19	15	0	0	0	0	"	"	"	"	"	3/8x4x32	"	8x10	"	"
	2x8	2x6	47	43	40	20	16	12	0	0	0	2x4	2x4	2x4	2x4	2x4	3/8x4x32	12x12	8x10	8x8	8x8
	2x10	4+4	68	63	60	29	26	17	0	0	0	"	"	"	"	"	½x4x28	14x16	12x10	"	10x8
	2x12	4+6	87	80	76	38	33	31	19	12	0	"	"	"	"	"	½x4x36	16x16	14x10	"	12x8
	2x12	6+6	97	89	90	42	38	35	21	16	12	"	"	"	"	"	½x4x44	18x16	16x10	"	14x8
4/12 Slope	2x4	2x4	30	28	27	13	0	0	0	0	0	2x4	2x4	2x4	2x4	2x4	3/8x3½x19	8x12	8x8	8x8	8x8
	2x6	2x4	42	39	34	18	0	0	0	0	0	"	"	"	"	"	3/8x4x23	10x12	"	"	"
	2x6	2x6	60	56	53	26	22	18	0	0	0	"	"	"	"	"	½x4x18	"	8x10	"	"
	2x8	2x6	64	59	57	28	24	19	0	0	0	2x4	2x4	2x4	2x4	2x4	½x4x25	14x16	12x10	8x8	10x8
	2x10	4+4	89	83	83	38	34	23	19	0	0	"	"	"	4&4	"	½x4x30	16x20	14x10	"	12x10
	2x12	4+6	100+	100+	100+	49	45	43	24	18	0	"	"	"	"	"	"	"	"	"	"
	2x12	6+6	-	-	-	54	50	48	27	23	18	"	"	"	"	"	½x4x34	"	16x12	"	14x10
5/12 Slope	2x4	2x4	33	32	30	14	0	0	0	0	0	2x4	2x4	2x4	2x4	2x4	3/8x3½x17	8x12	8x8	8x8	8x8
	2x6	2x4	52	48	43	22	0	0	0	0	0	"	"	"	"	"	3/8x4x22	10x12	"	"	"
	2x6	2x6	68	65	63	29	27	24	14	0	0	"	"	"	4&4	"	½x4x17	10x16	10x8	"	"
	2x8	2x6	79	74	71	34	31	26	17	0	0	2x4	2x4	2x4	4&4	2x4	½x4x18	12x16	10x8	8x8	8x8
	2x10	4+4	100+	100+	100+	47	43	32	23	0	0	"	"	2x6	"	"	½x4x25	14x20	14x12	8x10	10x8
	2x12	4+6	-	-	-	60	55	52	30	23	12	"	"	"	"	"	½x4x26	16x20	16x12	"	12x8
	2x12	6+6	-	-	-	65	60	57	32	28	23	"	"	"	"	"	½x4x30	18x16	18x12	"	14x8

See Ceiling Load Note, page 4.

1400f Lumber

Slope	Top chord	Bottom chord	Truss spacing, ft. 2' Ceiling dead load, psf 0	2' 5	2' 8	4' 0	4' 5	4' 8	8' 0	8' 5	8' 8	Web member sizes W1	W2	W3	W4	W5	Gusset Sizes, in. A T H W	B H W	C H W	D H W	E H W
			---Max. snow + roof dead load, psf---																		
3/12 Slope	2x4	2x4	32	29	28	0	0	0	0	0	0	2x4	2x4	2x4	2x4	2x4	3/8x3½x26	8x12	8x8	8x8	8x8
	2x6	2x4	41	39	36	0	12	0	0	0	0	"	"	"	"	"	½x4x16	10x12	"	"	"
	2x6	2x6	58	54	51	25	21	18	0	0	0	"	"	"	"	"	½x4x23	10x16	10x10	"	"
	2x8	2x6	61	57	54	26	22	19	0	0	0	2x4	2x4	2x4	2x4	2x4	½x4x23	12x16	10x10	8x8	8x8
	2x10	4+4	89	83	79	39	35	30	0	0	0	"	"	"	"	"	½x4x31	14x16	12x10	"	10x8
	2x12	4+6	100+	100+	100+	49	45	43	24	20	13	"	"	"	"	"	½x4x38	16x20	14x12	"	12x8
	2x12	6+6	–	–	–	52	47	45	26	21	18	"	"	"	"	"	½x4x44	18x20	16x12	"	14x8
4/12 Slope	2x4	2x4	37	35	33	16	14	0	0	0	0	2x4	2x4	2x4	2x4	2x4	3/8x3½x22	8x12	8x8	8x8	8x8
	2x6	2x4	55	52	49	24	18	0	0	0	0	"	"	"	"	"	½x4x16	10x12	8x10	"	"
	2x6	2x6	74	70	68	32	29	27	0	12	0	"	"	"	"	"	½x4x22	10x16	10x10	"	"
	2x8	2x6	84	77	74	36	32	30	0	12	0	2x4	2x4	2x4	2x4	2x4	½x4x23	12x16	10x10	8x8	8x8
	2x10	4+4	100+	100+	100+	50	47	40	25	15	0	"	"	"	"	"	½x4x32	14x20	12x12	"	10x10
	2x12	4+6	–	–	–	64	59	56	32	28	19	"	"	"	"	"	½x4x35	18x20	16x12	"	12x10
	2x12	6+6	–	–	–	66	61	58	33	29	26	"	"	"	"	"	½x4x38	"	18x12	"	14x10
5/12 Slope	2x4	2x4	41	39	38	17	16	0	0	0	0	2x4	2x4	2x4	2x4	2x4	3/8x3½x20	8x12	8x8	8x8	8x8
	2x6	2x4	68	64	58	30	23	0	0	0	0	"	"	"	"	"	½x4x16	10x16	10x8	"	"
	2x6	2x6	83	79	77	36	34	32	18	15	0	"	"	"	4&4	"	½x4x20	10x16	10x10	"	"
	2x8	2x6	100+	97	93	45	42	39	22	17	0	2x4	2x4	2x4	4&4	2x4	½x4x23	12x20	10x10	8x8	8x8
	2x10	4+4	–	100+	100+	62	58	49	31	20	0	"	"	2x6	"	"	½x4x32	16x20	14x14	8x12	10x10
	2x12	4+6	–	–	–	78	73	72	39	35	24	"	"	"	"	"	½x4x35	18x20	16x4	"	12x10
	2x12	6+6	–	–	–	79	73	70	39	35	33	"	"	"	"	"	½x4x37	18x20	18x14	"	14x10

1600f Lumber

Slope	Top chord	Bottom chord	Truss spacing, ft. 2' Ceiling dead load, psf 0	2' 5	2' 8	4' 0	4' 5	4' 8	8' 0	8' 5	8' 8	Web member sizes W1	W2	W3	W4	W5	Gusset Sizes, in. A T H W	B H W	C H W	D H W	E H W
			---Max. snow + roof dead load, psf---																		
3/12 Slope	2x4	2x4	38	37	35	16	14	0	0	0	0	2x4	2x4	2x4	2x4	2x4	½x3½x17	8x12	8x8	8x8	8x8
	2x6	2x4	49	46	45	21	18	0	0	0	0	"	"	"	"	"	½x4x19	10x12	8x10	"	"
	2x6	2x6	70	64	61	30	26	23	0	0	0	"	"	"	"	"	½x4x27	10x16	10x10	"	"
	2x8	2x6	74	68	65	32	28	25	0	0	0	2x4	2x4	2x4	2x4	2x4	½x4x27	12x16	10x10	8x8	8x8
	2x10	4+4	100+	99	100+	46	43	40	23	17	0	"	"	"	"	"	½x4x37	14x20	12x12	"	10x8
	2x12	4+6	–	–	–	59	54	52	29	25	20	"	"	"	"	"	½x4x46	18x20	16x12	"	12x8
	2x12	6+6				63	58	55	31	27	24	"	"	"	"	"	½x4x45	"	18x12	"	14x8
4/12 Slope	2x4	2x4	42	44	41	19	17	15	0	0	0	2x4	2x4	2x4	2x4	2x4	½x3½x15	8x12	8x8	8x8	8x8
	2x6	2x4	66	63	60	28	25	20	0	0	0	"	"	"	"	"	½x4x19	12x12	8x10	"	"
	2x6	2x6	90	85	82	39	37	34	19	16	12	"	"	"	"	"	½x4x26	10x16	10x10	"	8x10
	2x8	2x6	100+	95	91	44	41	39	22	18	16	2x4	2x4	2x4	2x4	2x4	½x4x28	12x16	10x10	8x8	8x10
	2x10	4+4	–	100+	100+	61	58	55	30	27	24	"	"	"	"	"	2-½x6x22	16x20	14x12	8x10	12x10
	2x12	4+6	–	–	–	77	73	70	38	35	33	"	"	"	"	"	2-½x6x28	18x20	16x12	"	14x10
	2x12	6+6	–	–	–	81	75	71	40	36	34	"	"	"	"	"	½x4x46	"	20x14	"	16x10
5/12 Slope	2x4	2x4	49	47	47	21	20	0	0	0	0	2x4	2x4	2x4	2x4	2x4	½x3½x13	8x12	8x8	8x8	8x8
	2x6	2x4	82	76	75	35	31	0	17	0	0	"	"	"	"	"	½x4x19	10x16	10x10	"	"
	2x6	2x6	100+	97	94	44	42	41	22	19	12	"	"	"	"	"	½x4x24	10x20	12x10	"	"
	2x8	2x6	–	100+	100+	54	50	48	27	23	12	2x4	2x4	2x4	2x4	2x4	2-½x6x14	14x16	12x10	8x8	8x8
	2x10	4+4	–	–		75	69	62	37	29	12	"	"	"	"	"	2-½x6x18	16x20	14x12	"	10x10
	2x12	4+6	–	–	–	94	87	87	47	43	35	"	"	2x6	"	"	2-½x6x22	20x24	18x14	8x10	12x10
	2x12	6+6	–	–	–	97	90	90	48	45	43	"	"	"	"	"	2-½x6x28	"	20x16	10x12	16x10

38' SPAN, 4-WEB

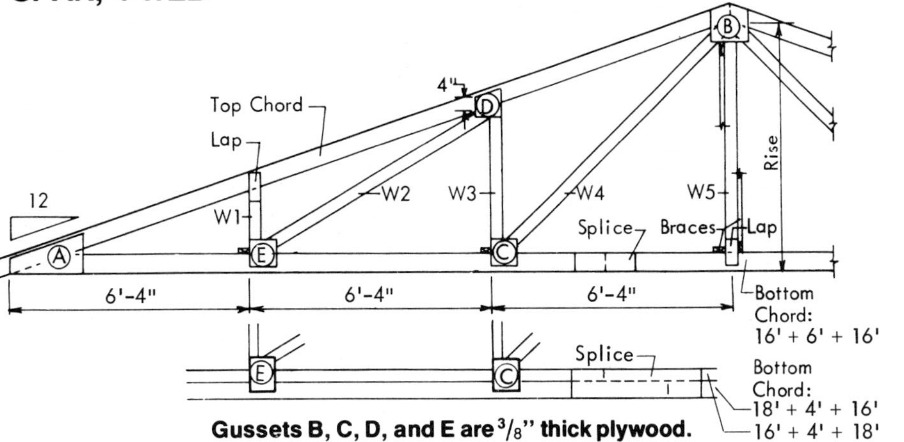

Gussets B, C, D, and E are $^3/_8$" thick plywood.

4+4, 4+6, 6+6 indicates stacked lower chord.
4&4, 6&4, indicate double web; a 2x4 is attached to the web member to increase its stiffness.

Before selecting heel gusset A, see pages 8 and 9 .

Web Lengths

Roof Slope	Rise	Top Chord	W1	W2	W3	W4	W5
3/12	4'-9"	20'	2'	7'	3'	8'	5'
4/12	6'-4"	16'+5'	2'	8'	4'	9'	6'
5/12	7'-11"	16'+5'	3'	8'	5'	10'+9'	8'

1100f Lumber

			Truss spacing, ft. — Ceiling dead load, psf — ---Max. snow + roof dead load, psf---									Web member sizes					Gusset Sizes, in.				
			2'			4'			8'								A	B	C	D	E
Slope	Top chord	Bottom chord	0	5	8	0	5	8	0	5	8	W1	W2	W3	W4	W5	T H W	H W	H W	H W	H W
3/12 Slope	2x4	2x4	24	22	19	0	0	0	0	0	0	2x4	2x4	2x4	2x4	2x4	3/8x3½x21	8x12	8x8	8x8	8x8
	2x6	2x4	30	27	20	0	0	0	0	0	0	"	"	"	"	"	3/8x4x22	10x12	"	"	"
	2x6	2x6	42	38	35	18	13	0	0	0	0	"	"	"	"	"	3/8x4x32	"	8x10	"	"
	2x8	2x6	44	41	38	19	15	0	0	0	0	2x4	2x4	2x4	2x4	2x4	3/8x4x32	12x12	8x10	8x8	8x8
	2x10	4+4	64	59	57	28	23	14	0	0	0	"	"	"	"	"	½x4x28	14x16	12x10	"	10x8
	2x12	4+6	82	75	72	35	31	28	0	0	0	"	"	"	"	"	½x4x36	16x16	14x10	"	12x8
	2x12	6+6	92	84	80	40	35	33	20	15	0	"	"	"	4&4	"	½x4x44	18x16	16x10	"	14x8
4/12 Slope	2x4	2x4	27	26	25	0	0	0	0	0	0	2x4	2x4	2x4	2x4	2x4	3/8x3½x19	8x12	8x8	8x8	8x8
	2x6	2x4	40	37	31	0	0	0	0	0	0	"	"	"	"	"	3/8x4x22	10x12	"	"	"
	2x6	2x6	56	53	50	24	20	15	0	0	0	"	"	"	"	"	½x4x18	12x12	8x10	"	"
	2x8	2x6	61	56	54	26	22	15	0	0	0	2x4	2x4	2x4	2x4	2x4	½x4x18	14x12	8x10	8x8	8x8
	2x10	4+4	84	78	78	36	31	20	0	0	0	"	"	"	"	"	½x4x25	14x16	12x10	"	10x8
	2x12	4+6	100 +	98	99	46	42	39	23	15	0	"	"	"	4&4	"	½x4x30	16x20	14x10	"	12x10
	2x12	6+6	–	100+	100+	51	47	45	26	22	14	"	"	"	"	"	½x4x35	18x16	18x12	"	14x10
5/12 Slope	2x4	2x4	30	29	28	13	0	0	0	0	0	2x4	2x4	2x4	2x4	2x4	3/8x3½x17	8x12	8x8	8x8	8x8
	2x6	2x4	49	46	38	21	0	0	0	0	0	"	"	"	"	"	3/8x4x22	10x12	"	"	"
	2x6	2x6	63	60	58	27	25	21	0	0	0	"	"	"	"	"	½x4x17	10x16	10x8	"	"
	2x8	2x6	75	70	67	33	29	21	0	0	0	2x4	2x4	2x4	2x4	2x4	½x4x18	12x16	10x8	8x8	8x8
	2x10	4+4	100+	100+	96	45	39	26	22	0	0	"	"	2x6	4&4	"	½x4x25	14x20	14x12	8x10	10x8
	2x12	4+6	–	–	100+	57	52	48	28	20	0	"	"	"	"	"	½x4x31	18x16	16x12	"	12x8
	2x12	6+6	–	–	–	62	57	54	31	27	19	"	"	"	"	"	½x4x31	"	18x12	8x12	14x10

1400f Lumber

Slope	Top chord	Bottom chord	2' Truss spacing, ft.: 0	2': 5	2': 8	4': 0	4': 5	4': 8	8': 0	8': 5	8': 8	W1	W2	W3	W4	W5	A T H W	B H W	C H W	D H W	E H W
			Ceiling dead load, psf / ---Max. snow + roof dead load, psf---									Web member sizes					Gusset Sizes, in.				
3/12 Slope	2x4	2x4	29	27	25	0	0	0	0	0	0	2x4	2x4	2x4	2x4	2x4	3/8x3½x25	8x12	8x8	8x8	8x8
	2x6	2x4	39	36	34	0	0	0	0	0	0	"	"	"	"	"	½x4x16	10x12	"	"	"
	2x6	2x6	56	51	48	24	19	16	0	0	0	"	"	"	"	"	½x4x23	10x16	10x10	"	"
	2x8	2x6	58	54	51	25	21	18	0	0	0	2x4	2x4	2x4	2x4	2x4	½x4x23	12x16	10x10	8x8	8x8
	2x10	4+4	84	78	75	36	33	26	0	0	0	"	"	"	"	"	½x4x31	14x16	12x10	"	10x8
	2x12	4+6	100 +	99	100	46	43	40	23	19	0	"	"	"	"	"	½x4x39	16x20	14x12	"	12x8
	2x12	6+6	–	–	–	49	45	42	24	20	17	"	"	"	"	"	½x4x44	18x20	18x12	"	14x8
4/12 Slope	2x4	2x4	34	32	31	14	13	0	0	0	0	2x4	2x4	2x4	2x4	2x4	3/8x3½x22	8x12	8x8	8x8	8x8
	2x6	2x4	52	49	46	22	13	0	0	0	0	"	"	"	"	"	½x4x16	10x12	8x10	"	"
	2x6	2x6	68	64	63	29	27	24	0	0	0	"	"	"	"	"	½x4x22	10x16	10x10	"	8x10
	2x8	2x6	79	74	70	34	31	28	0	0	0	2x4	2x4	2x4	2x4	2x4	½x4x23	12x16	10x10	8x8	8x8
	2x10	4+4	100+	100+	100+	48	45	35	24	13	0	"	"	"	4&4	"	½x4x32	14x20	12x12	"	10x10
	2x12	4+6	–	–	–	60	56	53	30	25	15	"	"	"	"	"	½x4x35	18x20	16x12	"	12x10
	2x12	6+6	–	–	–	63	58	55	31	27	25	"	"	"	"	"	½x4x39	18x20	18x12	8x10	14x10
5/12 Slope	2x4	2x4	37	36	35	16	14	0	0	0	0	2x4	2x4	2x4	2x4	2x4	3/8x3½x20	8x12	8x8	8x8	8x8
	2x6	2x4	65	60	55	28	17	0	0	0	0	"	"	"	"	"	½x4x16	10x16	10x8	"	"
	2x6	2x6	76	73	71	33	31	29	16	14	0	"	"	"	4&4	"	½x4x20	"	10x10	"	"
	2x8	2x6	99	92	88	43	40	35	21	14	0	2x4	2x4	2x4	4&4	2x4	½x4x24	12x20	10x10	8x8	8x8
	2x10	4+4	–	100+	100+	59	54	44	29	17	0	"	"	2x6	"	"	2–½x6x18	16x20	14x14	8x12	10x10
	2x12	4+6	–	–	–	74	69	67	37	32	20	"	"	"	"	"	½x4x35	18x20	16x14	"	12x10
	2x12	6+6	–	–	–	75	69	66	37	33	31	"	"	"	"	"	½x4x37	"	18x14	"	14x10

1600f Lumber

Slope	Top chord	Bottom chord	2' Truss spacing, ft.: 0	2': 5	2': 8	4': 0	4': 5	4': 8	8': 0	8': 5	8': 8	W1	W2	W3	W4	W5	A T H W	B H W	C H W	D H W	E H W
			Ceiling dead load, psf / ---Max. snow + roof dead load, psf---									Web member sizes					Gusset Sizes, in.				
3/12 Slope	2x4	2x4	35	34	32	15	13	0	0	0	0	2x4	2x4	2x4	2x4	2x4	½x3½x17	8x12	8x8	8x8	8x8
	2x6	2x4	47	44	42	20	15	0	0	0	0	"	"	"	"	"	½x4x20	10x16	8x10	"	"
	2x6	2x6	67	61	58	29	24	21	0	0	0	"	"	"	"	"	½x4x27	"	10x10	"	"
	2x8	2x6	70	64	61	30	26	23	0	0	0	2x4	2x4	2x4	2x4	2x4	½x4x28	12x16	10x10	8x8	8x8
	2x10	4+4	100+	94	90	44	41	35	22	14	0	"	"	"	"	"	½x4x37	14x20	12x12	"	10x8
	2x12	4+6	–	100+	100+	56	51	49	28	24	16	"	"	"	"	"	½x4x46	18x20	16x12	"	12x8
	2x12	6+6	–	–	–	60	55	52	30	25	23	"	"	"	"	"	½x4x46	"	18x12	8x10	14x8
4/12 Slope	2x4	2x4	41	39	37	17	15	14	0	0	0	2x4	2x4	2x4	2x4	2x4	½x3½x15	8x12	8x8	8x8	8x8
	2x6	2x4	62	60	56	27	23	18	0	0	0	"	"	"	"	"	½x4x20	12x12	8x10	8x8	8x8
	2x6	2x6	83	78	76	36	33	31	18	15	0	"	"	"	"	"	½x4x26	10x16	10x10	"	8x10
	2x8	2x6	96	90	87	41	39	36	20	17	14	2x4	2x4	2x4	2x4	2x4	½x4x28	12x20	10x10	8x8	8x10
	2x10	4+4	100+	100+	100+	58	55	52	29	25	22	"	"	"	"	"	2–½x6x22	16x20	14x12	8x10	12x10
	2x12	4+6	–	–	–	73	69	66	36	33	31	"	"	"	"	"	2–½x6x28	18x20	16x12	"	14x10
	2x12	6+6	–	–	–	77	71	67	38	34	32	"	"	"	"	"	2–½x6x34	"	20x14	"	16x10
5/12 Slope	2x4	2x4	45	44	43	19	18	0	0	0	0	2x4	2x4	2x4	2x4	2x4	½x3½x13	8x12	8x8	8x8	8x8
	2x6	2x4	78	72	70	34	27	0	0	0	0	"	"	"	"	"	½x4x19	10x16	10x10	"	"
	2x6	2x6	93	89	87	40	39	37	20	18	0	"	"	"	"	"	½x4x24	10x20	12x10	"	"
	2x8	2x6	100+	100+	100+	51	48	46	25	20	0	2x4	2x4	2x4	2x4	2x4	2–½x6x15	14x16	12x10	8x10	8x10
	2x10	4+4	–	–		71	68	54	35	24	0	"	"	2x6	"	"	2–½x6x18	16x20	14x14	8x12	10x10
	2x12	4+6	–	–	–	89	83	82	44	41	30	"	"	"	"	"	2–½x6x22	20x24	18x14	"	12x10
	2x12	6+6	–	–	–	92	85	81	46	42	39	"	"	"	"	"	2–½x6x28	"	20x16	10x12	16x10

40' SPAN, 4-WEB

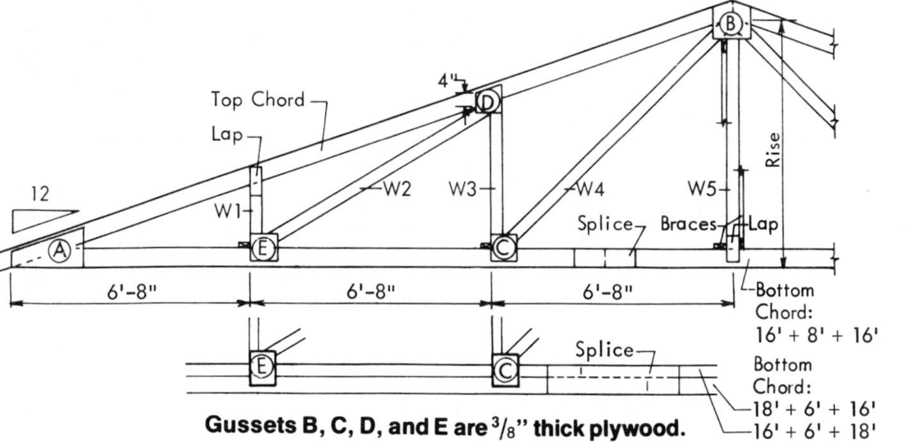

Gussets B, C, D, and E are 3/8" thick plywood.

4+4, 4+6, 6+6 indicates stacked lower chord.
4&4, 6&4, indicate double web; a 2x4 is attached to the web member to increase its stiffness.

Before selecting heel gusset A, see pages 8 and 9 .

Web Lengths

Roof Slope	Rise	Top Chord	W1	W2	W3	W4	W5
3/12	5'-0"	16'+5'	2'	7'	3'	8'	5'
4/12	6'-8"	18'+4'	2'	8'	5'	9'+8'	7'
5/12	8'-4"	18'+4'	3'	9'	6'	11'+10'	8'

1100f Lumber

See Ceiling Load Note, page 4.

			Truss spacing, ft. 2'			4'			8'			Web member sizes					Gusset Sizes, in. A	B	C	D	E
			Ceiling dead load, psf																		
Slope	Top chord	Bottom chord	0	5	8	0	5	8	0	5	8	W1	W2	W3	W4	W5	T H W	H W	H W	H W	H W
			---Max. snow + roof dead load, psf---																		
3/12 Slope	2x4	2x4	22	20	16	0	0	0	0	0	0	2x4	2x4	2x4	2x4	2x4	3/8x3½x21	8x12	8x8	8x8	8x8
	2x6	2x4	28	25	15	0	0	0	0	0	0	"	"	"	"	"	3/8x4x22	10x12	"	"	"
	2x6	2x6	41	36	33	0	13	0	0	0	0	"	"	"	"	"	3/8x4x32	"	8x10	"	"
	2x8	2x6	42	38	35	0	14	0	0	0	0	2x4	2x4	2x4	2x4	2x4	3/8x4x33	12x12	8x10	8x8	8x8
	2x10	4+4	61	56	54	26	20	0	0	0	0	"	"	"	"	"	½x4x28	14x16	12x10	"	10x8
	2x12	4+6	77	71	68	33	29	25	0	0	0	"	"	"	"	"	½x4x36	16x16	14x10	"	12x8
	2x12	6+6	87	80	76	38	33	31	0	14	0	"	"	"	"	"	½x4x44	18x16	16x10	"	14x8
4/12 Slope	2x4	2x4	25	24	23	0	0	0	0	0	0	2x4	2x4	2x4	2x4	2x4	3/8x3½x19	8x12	8x8	8x8	8x8
	2x6	2x4	38	34	23	0	0	0	0	0	0	"	"	"	"	"	3/8x4x22	10x12	"	"	"
	2x6	2x6	51	49	47	22	18	12	0	0	0	"	"	"	"	"	½x4x18	12x12	8x10	"	"
	2x8	2x6	58	53	51	25	21	13	0	0	0	2x4	2x4	2x4	2x4	2x4	½x4x19	14x12	8x10	8x8	8x8
	2x10	4+4	80	74	72	34	28	14	0	0	0	"	"	"	"	"	½x4x25	14x16	12x10	"	10x8
	2x12	4+6	100+	93	94	44	40	34	0	13	0	"	"	"	"	"	½x4x31	16x20	14x12	"	12x10
	2x12	6+6	-	100+	100+	49	45	43	24	20	0	"	"	"	4&4	"	½x4x35	18x16	18x12	8x10	14x10
5/12 Slope	2x4	2x4	28	26	26	0	0	0	0	0	0	2x4	2x4	2x4	2x4	2x4	3/8x3½x17	8x12	8x8	8x8	8x8
	2x6	2x4	47	42	33	0	0	0	0	0	0	"	"	"	"	"	3/8x4x22	10x12	"	"	"
	2x6	2x6	58	55	54	25	23	17	0	0	0	"	"	"	"	"	½x4x17	10x16	8x10	"	"
	2x8	2x6	72	67	64	31	27	17	0	0	0	2x4	2x4	2x4	2x4	2x4	½x4x18	12x16	10x10	8x8	8x8
	2x10	4+4	99	96	88	43	35	21	0	0	0	"	"	2x6	"	"	½x4x25	14x20	14x12	8x10	10x8
	2x12	4+6	-	100+	100+	54	50	43	27	17	0	"	"	2x8	4&4	"	½x4x31	18x16	16x14	8x14	12x10
	2x12	6+6	-	-	-	58	54	52	29	25	16	"	"	"	"	"	½x4x31	18x20	18x14	"	14x10

1400f Lumber

Slope	Top chord	Bottom chord	Truss spacing, ft. 2' — Ceiling dead load, psf 0	2' 5	2' 8	4' 0	4' 5	4' 8	8' 0	8' 5	8' 8	Web member sizes W1	W2	W3	W4	W5	Gusset Sizes, in. A T H W	B H W	C H W	D H W	E H W
			---Max. snow + roof dead load, psf---																		
3/12 Slope	2x4	2x4	27	25	23	0	0	0	0	0	0	2x4	2x4	2x4	2x4	2x4	3/8x3½x25	8x12	8x8	8x8	8x8
	2x6	2x4	37	34	32	0	0	0	0	0	0	"	"	"	"	"	½x4x16	10x12	"	"	"
	2x6	2x6	53	49	46	23	18	15	0	0	0	"	"	"	"	"	½x4x23	10x16	10x10	"	"
	2x8	2x6	55	51	49	24	19	16	0	0	0	2x4	2x4	2x4	2x4	2x4	½x4x24	12x16	10x10	8x8	8x8
	2x10	4+4	80	74	71	34	31	22	0	0	0	"	"	"	"	"	½x4x31	14x16	12x10	"	10x8
	2x12	4+6	100+	93	94	44	40	37	0	17	0	"	"	"	"	"	½x4x39	16x20	14x12	"	12x8
	2x12	6+6	–	98	99	46	42	39	23	19	16	"	"	"	4&4	"	½x4x44	18x20	18x12	8x10	14x8
4/12 Slope	2x4	2x4	31	29	28	0	0	0	0	0	0	2x4	2x4	2x4	2x4	2x4	3/8x3½x22	8x12	8x8	8x8	8x8
	2x6	2x4	49	47	43	0	0	0	0	0	0	"	"	"	"	"	½x4x17	10x12	8x10	"	"
	2x6	2x6	63	60	58	27	25	23	0	0	0	"	"	"	"	"	½x4x22	10x16	10x10	"	8x10
	2x8	2x6	75	70	67	33	29	25	0	0	0	2x4	2x4	2x4	2x4	2x4	½x4x23	12x16	10x10	8x8	8x8
	2x10	4+4	100+	98	98	45	41	31	0	0	0	"	"	"	"	"	½x4x33	14x20	12x12	"	10x10
	2x12	4+6	–	–	–	57	53	50	28	23	12	"	"	"	4&4	"	½x4x40	18x20	16x12	8x10	14x10
	2x12	6+6	–	–	–	60	55	52	30	25	23	"	"	"	"	"	½x4x38	18x20	18x12	"	16x10
5/12 Slope	2x4	2x4	34	33	32	15	13	0	0	0	0	2x4	2x4	2x4	2x4	2x4	3/8x3½x20	8x12	8x8	8x8	8x8
	2x6	2x4	62	57	52	27	13	0	0	0	0	"	"	"	"	"	½x4x16	10x16	10x8	"	"
	2x6	2x6	71	68	66	31	28	27	0	0	0	"	"	"	"	"	½x4x20	"	10x10	"	"
	2x8	2x6	94	88	84	41	37	32	0	12	0	2x4	2x4	2x6	2x4	2x4	½x4x24	12x20	10x12	8x10	8x8
	2x10	4+4	100+	100+	100+	56	51	39	28	14	0	"	"	"	4&4	"	2–½x6x18	16x20	14x14	8x12	10x10
	2x12	4+6	–	–	–	70	65	63	35	29	16	"	"	"	"	"	½x4x35	18x20	16x14	"	12x10
	2x12	6+6	–	–	–	71	66	63	35	31	29	"	"	"	"	"	½x4x38	20x24	20x14	"	14x10

1600f Lumber

Slope	Top chord	Bottom chord	Truss spacing, ft. 2' — Ceiling dead load, psf 0	2' 5	2' 8	4' 0	4' 5	4' 8	8' 0	8' 5	8' 8	Web member sizes W1	W2	W3	W4	W5	Gusset Sizes, in. A T H W	B H W	C H W	D H W	E H W
			---Max. snow + roof dead load, psf---																		
3/12 Slope	2x4	2x4	33	31	29	0	12	0	0	0	0	2x4	2x4	2x4	2x4	2x4	½x3½x17	8x12	8x8	8x8	8x8
	2x6	2x4	44	42	39	0	0	0	0	0	0	"	"	"	"	"	½x4x19	10x16	8x10	"	"
	2x6	2x6	64	58	55	27	23	20	0	0	0	"	"	"	"	"	½x4x27	"	10x10	"	"
	2x8	2x6	67	61	58	29	25	21	0	0	0	2x4	2x4	2x4	2x4	2x4	½x4x28	12x16	10x10	8x8	8x8
	2x10	4+4	96	89	89	41	39	31	0	12	0	"	"	"	"	"	½x4x38	16x20	12x12	8x10	10x8
	2x12	4+6	100+	100+	100+	53	48	46	26	22	13	"	"	"	"	"	½x4x46	18x20	16x12	"	12x8
	2x12	6+6	–	–	–	57	52	49	28	24	21	"	"	"	"	"	½x4x47	"	18x12	"	14x10
4/12 Slope	2x4	2x4	38	36	34	16	14	12	0	0	0	2x4	2x4	2x4	2x4	2x4	½x3½x14	8x12	8x8	8x8	8x8
	2x6	2x4	59	57	53	26	21	16	0	0	0	"	"	"	"	"	½x4x20	12x12	8x10	"	"
	2x6	2x6	77	72	70	33	30	29	0	13	0	"	"	"	"	"	½x4x26	10x16	10x10	"	8x10
	2x8	2x6	91	86	83	39	37	34	0	16	13	2x4	2x4	2x4	2x4	2x4	½x4x28	12x20	10x10	8x8	8x10
	2x10	4+4	100+	100+	100+	55	52	49	27	24	20	"	"	"	"	"	2–½x6x22	16x20	14x12	8x10	12x10
	2x12	4+6	–	–	–	69	66	63	34	31	29	"	"	"	"	"	2–½x6x28	18x20	16x12	"	14x10
	2x12	6+6	–	–	–	73	67	64	36	32	30	"	"	"	"	"	2–½x6x34	18x20	20x14	"	16x10
5/12 Slope	2x4	2x4	41	40	40	18	16	0	0	0	0	2x4	2x4	2x4	2x4	2x4	½x3½x13	8x12	8x8	8x8	8x8
	2x6	2x4	75	69	65	32	24	0	0	0	0	"	"	"	"	"	½x4x19	10x16	10x10	"	"
	2x6	2x6	86	82	80	37	36	34	18	16	0	"	"	"	4&4	"	½x4x23	10x20	12x10	"	"
	2x8	2x6	100+	100+	100+	49	46	43	24	16	0	2x4	2x4	2x6	4&4	2x4	½x6x15	14x16	12x12	8x12	8x10
	2x10	4+4	–	–		68	64	51	34	21	0	"	"	"	"	"	2–½x6x18	16x20	14x14	"	10x10
	2x12	4+6	–	–	–	85	79	78	42	38	27	"	"	"	"	"	2–½x6x22	20x24	18x14	"	12x10
	2x12	6+6	–	–	–	87	81	77	43	40	31	"	"	"	"	"	2–½x6x28	"	20x16	8x14	14x12

42' SPAN, 4-WEB

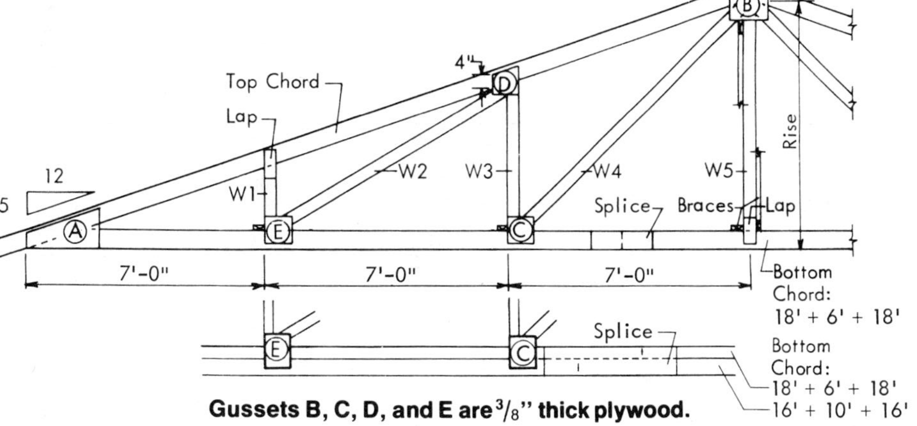

Gussets B, C, D, and E are $^3/_8$" thick plywood.

4+4, 4+6, 6+6 indicates stacked lower chord.
4&4, 6&4, indicate double web; a 2x4 is attached to the web member to increase its stiffness.

Before selecting heel gusset A, see pages 8 and 9 .

Web Lengths

Roof Slope	Rise	Top Chord	W1	W2	W3	W4	W5
3/12	5'-3"	18'+4'	2'	8'	4'	9'+8'	5'
4/12	7'-0"	18'+5'	2'	8'	5'	10'+9'	7'
5/12	8'-9"	18'+5'	3'	9'	6'	11'	9'

1100f Lumber

See Ceiling Load Note, page 4.

			Truss spacing, ft.									Web member sizes					Gusset Sizes, in.				
			2'			4'			8'								A	B	C	D	E
			Ceiling dead load, psf																		
	Top chord	Bottom chord	0	5	8	0	5	8	0	5	8	W1	W2	W3	W4	W5	T H W	H W	H W	H W	H W
			---Max. snow + roof dead load, psf---																		
3/12 Slope	2x4	2x4	20	18	0	0	0	0	0	0	0	2x4	2x4	2x4	2x4	2x4	3/8x3½x20	8x12	8x8	8x8	8x8
	2x6	2x4	27	24	0	0	0	0	0	0	0	"	"	"	"	"	3/8x4x23	10x12	"	"	"
	2x6	2x6	39	34	31	0	12	0	0	0	0	"	"	"	"	"	3/8x4x32	"	8x10	"	"
	2x8	2x6	40	36	33	0	13	0	0	0	0	2x4	2x4	2x4	2x4	2x4	3/8x4x33	12x12	8x10	8x8	8x8
	2x10	4+4	58	54	51	25	18	0	0	0	0	"	"	"	"	"	½x4x28	14x16	12x10	"	10x8
	2x12	4+6	73	67	64	32	28	22	0	0	0	"	"	"	"	"	½x4x36	16x16	14x10	"	12x8
	2x12	6+6	83	76	72	36	31	29	0	13	0	"	"	"	"	"	½x4x44	18x16	16x12	"	14x8
4/12 Slope	2x4	2x4	23	22	19	0	0	0	0	0	0	2x4	2x4	2x4	2x4	2x4	3/8x3½x18	8x12	8x8	8x8	8x8
	2x6	2x4	36	32	17	0	0	0	0	0	0	"	"	"	"	"	3/8x4x22	10x12	"	"	"
	2x6	2x6	48	46	44	21	17	0	0	0	0	"	"	"	"	"	½x4x18	12x12	8x10	"	"
	2x8	2x6	55	51	49	24	19	0	0	0	0	2x4	2x4	2x4	2x4	2x4	½x4x19	14x12	8x10	8x8	8x8
	2x10	4+4	76	71	67	33	25	0	0	0	0	"	"	"	"	"	½x4x25	14x16	12x10	"	10x10
	2x12	4+6	96	89	89	41	37	31	0	0	0	"	"	"	"	"	½x4x30	16x20	14x12	"	12x10
	2x12	6+6	100+	99	100	46	43	40	23	18	0	"	4&4	"	4&4	"	½x4x35	18x20	18x12	8x10	16x10
5/12 Slope	2x4	2x4	26	24	24	0	0	0	0	0	0	2x4	2x4	2x4	2x4	2x4	3/8x3½x16	8x12	8x8	8x8	8x8
	2x6	2x4	45	40	23	0	0	0	0	0	0	"	"	"	"	"	3/8x4x22	10x12	"	"	"
	2x6	2x6	54	51	50	23	21	14	0	0	0	"	"	"	"	"	½x4x16	10x16	10x8	"	"
	2x8	2x6	69	64	61	30	25	14	0	0	0	"	"	2x6	"	"	½x4x19	12x16	10x12	8x10	8x8
	2x10	4+4	94	91	81	41	31	13	0	0	0	"	"	"	"	"	½x4x25	14x20	14x12	"	10x8
	2x12	4+6	100+	100+	100+	51	47	38	25	15	0	"	4&4	2x8	4&4	"	½x4x31	18x16	16x14	8x14	12x10
	2x12	6+6	-	-	-	56	51	49	28	24	12	"	"	"	"	"	½x4x31	18x20	18x14	"	14x10

1400f Lumber

Slope	Top chord	Bottom chord	Truss spacing 2', Ceiling dead load 0 psf	2', 5	2', 8	4', 0	4', 5	4', 8	8', 0	8', 5	8', 8	W1	W2	W3	W4	W5	A T H W	B H W	C H W	D H W	E H W
			---Max. snow + roof dead load, psf---									Web member sizes					Gusset Sizes, in.				
3/12 Slope	2x4	2x4	25	23	22	0	0	0	0	0	0	2x4	2x4	2x4	2x4	2x4	3/8x3½x25	8x12	8x8	8x8	8x8
	2x6	2x4	35	32	28	0	0	0	0	0	0	"	"	"	"	"	½x4x16	10x12	"	"	"
	2x6	2x6	50	47	44	0	17	14	0	0	0	"	"	"	"	"	½x4x24	10x16	10x10	"	"
	2x8	2x6	53	49	46	0	18	14	0	0	0	2x4	2x4	2x4	2x4	2x4	½x4x24	12x16	10x10	8x8	8x8
	2x10	4+4	76	70	67	33	29	19	0	0	0	"	"	"	"	"	½x4x32	16x16	12x10	"	10x8
	2x12	4+6	96	88	89	42	37	35	0	14	0	"	"	"	"	"	½x4x39	16x20	14x12	"	12x8
	2x12	6+6	100+	93	94	44	40	37	0	17	13	"	"	"	"	"	½x4x44	18x20	18x12	8x10	14x10
4/12 Slope	2x4	2x4	29	27	26	0	0	0	0	0	0	2x4	2x4	2x4	2x4	2x4	3/8x3½x22	8x12	8x8	8x8	8x8
	2x6	2x4	47	45	39	0	0	0	0	0	0	"	"	"	"	"	½x4x17	10x12	8x10	"	"
	2x6	2x6	58	56	54	25	23	21	0	0	0	"	"	"	"	"	½x4x22	10x16	10x10	"	8x10
	2x8	2x6	72	67	64	31	27	21	0	0	0	2x4	2x4	2x4	2x4	2x4	½x4x24	12x16	10x10	8x8	8x10
	2x10	4+4	100	93	93	43	38	26	0	0	0	"	"	"	"	"	½x4x33	14x20	12x12	8x10	10x10
	2x12	4+6	-	100+	100+	54	50	48	27	20	0	"	"	"	4&4	"	½x4x40	18x20	16x12	"	14x10
	2x12	6+6	-	-	-	57	52	50	28	24	19	"	"	"	"	"	½x4x38	"	18x12	"	16x10
5/12 Slope	2x4	2x4	32	31	29	14	0	0	0	0	0	2x4	2x4	2x4	2x4	2x4	3/8x3½x19	8x12	8x8	8x8	8x8
	2x6	2x4	59	54	48	26	0	0	0	0	0	"	"	"	"	"	½x4x16	10x16	10x8	"	"
	2x6	2x6	66	63	61	28	26	25	0	0	0	"	"	"	"	"	½x4x19	"	10x10	"	"
	2x8	2x6	90	84	80	39	35	28	0	0	0	2x4	2x4	2x6	2x4	2x4	½x4x24	12x20	12x12	8x10	8x8
	2x10	4+4	100+	100+	100+	54	48	34	0	0	0	"	"	2x8	"	"	2-½x6x18	16x20	14x16	8x14	10x10
	2x12	4+6	-	-	-	67	62	55	33	26	13	"	"	"	4&4	"	½x4x35	18x20	16x16	"	12x10
	2x12	6+6	-	-	-	68	62	60	34	30	26	"	"	"	"	"	½x4x37	"	18x16	"	14x10

1600f Lumber

Slope	Top chord	Bottom chord	Truss spacing 2', Ceiling dead load 0 psf	2', 5	2', 8	4', 0	4', 5	4', 8	8', 0	8', 5	8', 8	W1	W2	W3	W4	W5	A T H W	B H W	C H W	D H W	E H W
			---Max. snow + roof dead load, psf---									Web member sizes					Gusset Sizes, in.				
3/12 Slope	2x4	2x4	30	28	27	0	0	0	0	0	0	2x4	2x4	2x4	2x4	2x4	½x3½x17	8x12	8x8	8x8	8x8
	2x6	2x4	42	40	37	0	0	0	0	0	0	"	"	"	"	"	½x4x20	10x16	8x10	"	"
	2x6	2x6	61	56	53	26	22	18	0	0	0	"	"	"	"	"	½x4x28	"	10x10	"	"
	2x8	2x6	64	59	56	27	23	20	0	0	0	2x4	2x4	2x4	2x4	2x4	½x4x28	12x16	10x10	8x8	8x8
	2x10	4+4	91	85	85	39	36	28	0	0	0	"	"	"	"	"	½x4x39	16x20	12x12	8x10	10x10
	2x12	4+6	100+	100+	100+	50	46	44	0	20	0	"	"	"	"	"	½x4x47	18x20	16x12	"	12x10
	2x12	6+6	-	-	-	54	49	47	27	23	20	"	"	"	"	"	½x4x47	20x20	18x14	"	14x10
4/12 Slope	2x4	2x4	35	33	31	15	13	0	0	0	0	2x4	2x4	2x4	2x4	2x4	½x3½x14	8x12	8x8	8x8	8x8
	2x6	2x4	57	54	50	24	19	14	0	0	0	"	"	"	"	"	½x4x20	12x12	8x10	"	"
	2x6	2x6	71	67	65	31	28	26	0	12	0	"	"	"	"	"	½x4x25	10x16	10x10	"	8x10
	2x8	2x6	87	82	79	38	35	32	0	15	0	2x4	2x4	2x4	2x4	2x4	½x4x29	12x20	10x10	8x8	8x10
	2x10	4+4	100+	100+	100+	52	49	47	26	22	18	"	"	"	4&4	"	2-½x6x22	16x20	14x12	8x10	12x10
	2x12	4+6	-	-	-	66	62	60	33	30	26	"	"	"	"	"	2-½x6x28	18x20	16x12	"	14x10
	2x12	6+6	-	-	-	69	64	61	34	30	28	"	"	"	"	"	2-½x6x34	18x22	20x14	"	16x10
5/12 Slope	2x4	2x4	38	37	37	16	15	0	0	0	0	2x4	2x4	2x4	4&4	2x4	½x3½x13	8x12	8x8	8x8	8x8
	2x6	2x4	71	65	61	31	17	0	0	0	0	"	"	"	"	"	½x4x19	10x16	10x10	"	"
	2x6	2x6	80	77	75	34	33	31	17	15	0	"	"	"	"	"	½x4x23	10x20	12x10	"	"
	2x8	2x6	100+	100+	100+	47	44	38	23	15	0	2x4	2x4	2x6	4&4	2x4	2-½x6x15	14x16	12x12	8x12	8x10
	2x10	4+4	-	-	-	65	57	47	32	18	0	"	"	"	"	"	2-½x6x18	16x20	14x14	"	10x10
	2x12	4+6	-	-	-	81	75	73	40	34	21	"	"	2x8	"	"	2-½x6x22	20x24	18x16	10x14	14x10
	2x12	6+6	-	-	-	83	77	73	41	37	35	"	"	"	"	"	2-½x6x28	"	20x18	8x16	14x12

44' SPAN, 4-WEB

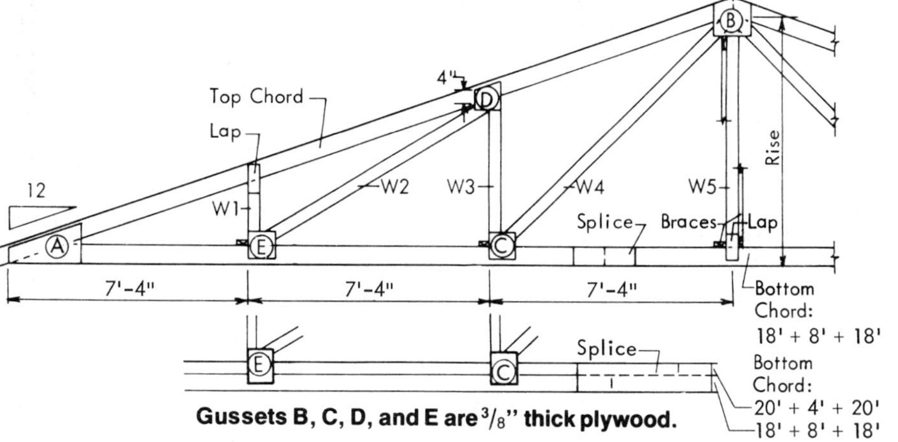

Gussets B, C, D, and E are 3/8" thick plywood.

4+4, 4+6, 6+6 indicates stacked lower chord.
4&4, 6&4, indicate double web; a 2x4 is attached to the web member to increase its stiffness.

Before selecting heel gusset A, see pages 8 and 9 .

Web Lengths

Roof Slope	Rise	Top Chord	W1	W2	W3	W4	W5
3/12	5'-6"	18'+5'	2'	8'	4'	9'+8'	6'
4/12	7'-4"	20'+4'	3'	9'	5'	10'+9'	7'
5/12	9'-2"	20'+4'	3'	10'	6'+5'	12'+11'	9'

1100f Lumber

See Ceiling Load Note, page 4.

			Truss spacing, ft.									Web member sizes					Gusset Sizes, in.				
			2'			4'			8'												
			Ceiling dead load, psf														A	B	C	D	E
	Top chord	Bottom chord	0	5	8	0	5	8	0	5	8	W1	W2	W3	W4	W5	T H W	H W	H W	H W	H W
			---Max. snow + roof dead load, psf---																		
3/12 Slope	2x4	2x4	0	17	0	0	0	0	0	0	0	2x4	2x4	2x4	2x4	2x4	3/8x3½x20	8x12	8x8	8x8	8x8
	2x6	2x4	0	22	0	0	0	0	0	0	0	"	"	"	"	"	3/8x4x23	10x12	"	"	"
	2x6	2x6	37	32	29	0	0	0	0	0	0	"	"	"	"	"	3/8x4x32	"	8x10	"	"
	2x8	2x6	39	34	31	0	12	0	0	0	0	2x4	2x4	2x4	2x4	2x4	3/8x4x33	12x12	8x10	8x8	8x8
	2x10	4+4	55	51	48	0	16	0	0	0	0	"	"	"	"	"	½x4x28	14x16	12x10	"	10x8
	2x12	4+6	70	64	61	30	26	18	0	0	0	"	"	"	"	"	½x4x36	16x16	14x10	"	12x8
	2x12	6+6	79	72	69	34	30	27	0	0	0	"	"	"	"	"	½x4x44	18x16	16x12	"	14x8
4/12 Slope	2x4	2x4	22	20	0	0	0	0	0	0	0	2x4	2x4	2x4	2x4	2x4	3/8x3½x17	8x12	8x8	8x8	8x8
	2x6	2x4	34	30	0	0	0	0	0	0	0	"	"	"	"	"	3/8x4x22	10x12	"	"	"
	2x6	2x6	44	42	41	0	16	0	0	0	0	"	"	"	"	"	½x4x18	12x12	8x10	"	"
	2x8	2x6	52	49	47	0	18	0	0	0	0	2x4	2x4	2x4	2x4	2x4	½x4x19	14x12	10x10	8x8	8x8
	2x10	4+4	72	67	61	31	22	0	0	0	0	"	"	"	"	"	½x4x25	14x16	12x10	"	10x10
	2x12	4+6	91	84	84	39	35	26	0	0	0	"	"	2x6	"	"	½x4x31	16x20	14x14	8x10	12x10
	2x12	6+6	100+	94	95	44	41	38	0	17	0	"	"	"	"	"	½x4x36	18x20	18x14	8x12	16x10
5/12 Slope	2x4	2x4	24	23	16	0	0	0	0	0	0	2x4	2x4	2x4	2x4	2x4	3/8x3½x15	8x12	8x8	8x8	8x8
	2x6	2x4	43	38	17	0	0	0	0	0	0	"	"	"	"	"	3/8x4x22	10x12	"	"	"
	2x6	2x6	50	48	47	21	19	0	0	0	0	"	"	"	4&4	"	½x4x16	10x16	10x8	"	"
	2x8	2x6	66	61	58	28	23	0	0	0	0	2x4	2x4	2x6	4&4	2x4	½x4x19	12x16	10x12	8x10	8x8
	2x10	4+4	90	86	75	39	28	0	0	0	0	"	"	2x8	"	"	½x4x26	14x20	14x14	8x12	10x8
	2x12	4+6	100+	100+	100+	49	45	34	0	12	0	"	"	"	"	"	½x4x31	18x16	16x14	8x14	12x10
	2x12	6+6	–	–	–	53	49	47	26	21	0	"	4&4	"	"	"	½x4x31	18x20	18x14	"	14x10

1400f Lumber

Slope	Top chord	Bottom chord	Truss spacing 2', ceiling dead load 0	2', 5	2', 8	4', 0	4', 5	4', 8	8', 0	8', 5	8', 8	W1	W2	W3	W4	W5	A T H W	B H W	C H W	D H W	E H W
			---Max. snow + roof dead load, psf---																		
3/12 Slope	2x4	2x4	23	21	20	0	0	0	0	0	0	2x4	2x4	2x4	2x4	2x4	3/8x3½x25	8x12	8x8	8x8	8x8
	2x6	2x4	34	31	22	0	0	0	0	0	0	"	"	"	"	"	½x4x17	10x12	"	"	"
	2x6	2x6	46	44	41	0	16	0	0	0	0	"	"	"	"	"	½x4x23	10x16	10x10	"	"
	2x8	2x6	51	47	44	0	17	12	0	0	0	2x4	2x4	2x4	2x4	2x4	½x4x25	12x16	10x10	8x8	8x8
	2x10	4+4	72	67	63	31	26	14	0	0	0	"	"	"	"	"	½x4x32	16x16	12x10	"	10x8
	2x12	4+6	91	84	80	40	35	31	0	12	0	"	"	"	"	"	½x4x39	16x20	14x12	"	12x8
	2x12	6+6	97	89	89	42	37	35	0	16	0	"	"	"	"	"	½x4x44	18x20	18x12	8x10	14x10
4/12 Slope	2x4	2x4	27	25	24	0	0	0	0	0	0	2x4	2x4	2x4	2x4	2x4	3/8x3½x22	8x12	8x8	8x8	8x8
	2x6	2x4	45	42	33	0	0	0	0	0	0	"	"	"	"	"	½x4x17	10x12	8x10	"	"
	2x6	2x6	54	52	50	23	21	17	0	0	0	"	"	"	"	"	½x4x21	12x12	10x10	"	8x10
	2x8	2x6	69	64	61	30	26	18	0	0	0	2x4	2x4	2x4	2x4	2x4	½x4x24	12x16	10x10	8x8	8x10
	2x10	4+4	95	88	88	41	35	23	0	0	0	"	"	"	"	"	½x4x33	14x20	12x12	8x10	10x10
	2x12	4+6	100+	100+	100+	52	48	43	0	17	0	"	"	2x6	"	"	½x4x40	18x20	16x14	8x12	14x10
	2x12	6+6	-	-	-	54	50	47	27	23	16	"	"	"	4&4	"	½x4x39	"	18x14	"	16x10
5/12 Slope	2x4	2x4	30	28	27	0	0	0	0	0	0	2x4	2x4	2x4	2x4	2x4	3/8x3½x19	8x12	8x8	8x8	8x8
	2x6	2x4	57	52	43	0	0	0	0	0	0	"	"	"	"	"	½x4x16	10x16	10x8	"	"
	2x6	2x6	61	58	57	26	24	23	0	0	0	"	"	"	"	"	½x4x19	"	10x10	"	"
	2x8	2x6	86	81	76	37	34	24	0	0	0	2x4	2x4	2x6	2x4	2x4	½x4x24	12x20	12x12	8x10	8x8
	2x10	4+4	100+	100+	100+	51	43	30	0	0	0	"	"	2x8	"	"	2-½x6x18	16x20	14x16	8x14	10x10
	2x12	4+6	-	-	-	64	59	52	32	23	0	"	4&4	"	4&4	"	½x4x35	18x20	16x16	"	12x10
	2x12	6+6	-	-	-	65	60	57	32	28	22	"	"	"	"	"	½x4x37	"	18x16	"	14x10

1600f Lumber

Slope	Top chord	Bottom chord	Truss spacing 2', ceiling dead load 0	2', 5	2', 8	4', 0	4', 5	4', 8	8', 0	8', 5	8', 8	W1	W2	W3	W4	W5	A T H W	B H W	C H W	D H W	E H W
			---Max. snow + roof dead load, psf---																		
3/12 Slope	2x4	2x4	28	26	25	0	0	0	0	0	0	2x4	2x4	2x4	2x4	2x4	½x3½x17	8x12	8x8	8x8	8x8
	2x6	2x4	40	38	35	0	0	0	0	0	0	"	"	"	"	"	½x4x20	10x16	8x10	"	"
	2x6	2x6	57	53	50	25	20	17	0	0	0	"	"	"	"	"	½x4x28	"	10x10	"	"
	2x8	2x6	61	56	53	26	22	18	0	0	0	2x4	2x4	2x4	2x4	2x4	½x4x29	12x16	10x10	8x8	8x8
	2x10	4+4	87	81	81	37	34	23	0	0	0	"	"	"	"	"	½x4x39	16x20	12x12	8x10	10x10
	2x12	4+6	100+	100+	100+	48	44	41	0	18	0	"	"	"	"	"	½x4x47	18x20	16x12	"	12x10
	2x12	6+6	-	-	-	51	47	45	25	21	17	"	"	"	4&4	"	½x4x48	20x20	18x14	"	14x10
4/12 Slope	2x4	2x4	32	30	29	0	12	0	0	0	0	2x4	2x4	2x4	2x4	2x4	½x3½x14	8x12	8x8	8x8	8x8
	2x6	2x4	54	52	48	0	18	12	0	0	0	"	"	"	"	"	½x4x20	12x12	8x10	"	"
	2x6	2x6	66	63	61	29	26	24	0	0	0	"	"	"	"	"	½x4x25	10x16	10x10	"	8x10
	2x8	2x6	83	79	75	36	33	30	0	14	0	2x4	2x4	2x4	2x4	2x4	½x4x29	12x20	10x10	8x8	8x10
	2x10	4+4	100+	100+	100+	50	47	44	0	20	15	"	"	"	"	"	2-½x6x22	16x20	14x12	8x10	12x10
	2x12	4+6	-	-	-	63	60	57	31	28	25	"	"	"	4&4	"	2-½x6x28	18x20	18x14	"	14x10
	2x12	6+6	-	-	-	66	61	58	33	29	27	"	"	"	"	"	2-½x6x34	18x22	20x14	10x10	16x12
5/12 Slope	2x4	2x4	36	35	34	15	14	0	0	0	0	2x4	2x4	2x4	4&4	2x4	3/8x3½x21	8x12	8x8	8x8	8x8
	2x6	2x4	68	62	54	30	13	0	0	0	0	"	"	"	"	"	½x4x20	10x16	10x10	"	"
	2x6	2x6	74	71	70	32	30	29	0	13	0	"	"	"	"	"	½x4x23	10x20	"	"	"
	2x8	2x6	100+	97	98	45	42	34	0	13	0	2x4	2x4	2x6	4&4	2x4	2-½x6x15	14x16	12x12	8x12	8x10
	2x10	4+4	-	100+	100+	62	53	41	31	14	0	"	"	2x8	"	"	2-½x6x18	16x20	14x16	8x14	10x10
	2x12	4+6	-	-	-	77	72	67	38	31	17	"	"	"	"	"	2-½x6x22	20x24	18x16	10x14	14x10
	2x12	6+6	-	-	-	79	73	70	39	35	32	"	"	"	"	"	2-½x6x28	"	20x18	8x16	14x12

46' SPAN, 4-WEB

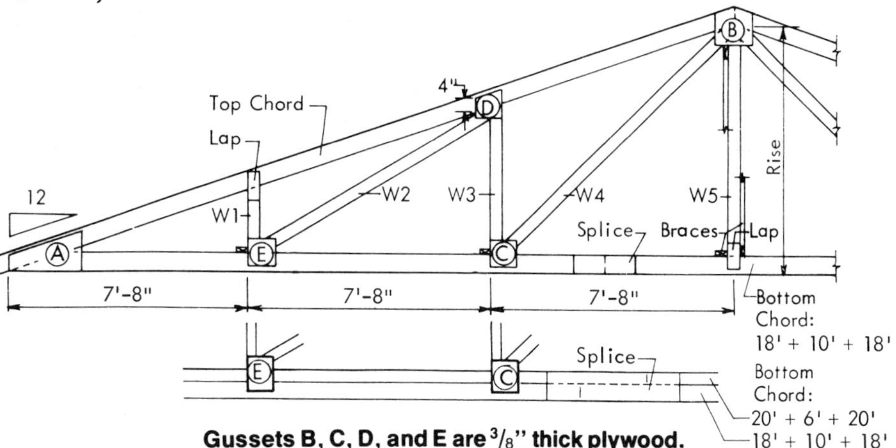

Gussets B, C, D, and E are 3/8" thick plywood.

4+4, 4+6, 6+6 indicates stacked lower chord.
4&4, 6&4, indicate double web; a 2x4 is attached to the web member to increase its stiffness.

Before selecting heel gusset A, see pages 8 and 9 .

Web Lengths

Roof Slope	Rise	Top Chord	W1	W2	W3	W4	W5
3/12	5'-9"	20'+4'	2'	9'	4'	10'+9'	6'
4/12	7'-8"	20'+5'	3'	9'	5'	11'+10'	8'
5/12	9'-7"	20'+6'	3'	10'+9'	6'+5'	12'+11'	10'

1100f Lumber

See Ceiling Load Note, page 4.

			Truss spacing, ft.									Web member sizes					Gussett Sizes, in.				
			2'			4'			8'								A	B	C	D	E
			Ceiling dead load, psf																		
	Top chord	Bottom chord	0	5	8	0	5	8	0	5	8	W1	W2	W3	W4	W5	T H W	H W	H W	H W	H W
			---Max. snow + roof dead load, psf---																		
3/12 Slope	2x4	2x4	0	16	0	0	0	0	0	0	0	2x4	2x4	2x4	2x4	2x4	3/8x3½x20	8x12	8x8	8x8	8x8
	2x6	2x4	0	20	0	0	0	0	0	0	0	"	"	"	"	"	3/8x4x22	10x12	"	"	"
	2x6	2x6	36	31	27	0	0	0	0	0	0	"	"	"	"	"	3/8x4x32	"	8x10	"	"
	2x8	2x6	37	33	30	0	0	0	0	0	0	2x4	2x4	2x4	2x4	2x4	3/8x4x34	12x12	8x10	8x8	8x8
	2x10	4+4	52	49	44	0	13	0	0	0	0	"	"	"	"	"	½x4x28	14x16	12x10	"	10x8
	2x12	4+6	66	61	58	0	25	16	0	0	0	"	"	"	"	"	½x4x36	16x16	14x10	"	12x8
	2x12	6+6	76	69	66	33	28	25	0	0	0	"	"	"	"	"	½x4x44	18x16	16x12	8x10	14x8
4/12 Slope	2x4	2x4	20	18	0	0	0	0	0	0	0	2x4	2x4	2x4	2x4	2x4	3/8x3½x17	8x12	8x8	8x8	8x8
	2x6	2x4	33	28	0	0	0	0	0	0	0	"	"	"	"	"	3/8x4x22	10x12	"	"	"
	2x6	2x6	42	40	39	0	15	0	0	0	0	"	"	"	"	"	½x4x18	12x12	8x10	"	"
	2x8	2x6	50	47	45	0	15	0	0	0	0	2x4	2x4	2x4	2x4	2x4	½x4x19	14x12	10x10	8x8	8x8
	2x10	4+4	69	64	55	30	19	12	0	0	0	"	"	"	4&4	"	½x4x25	14x16	12x10	"	10x10
	2x12	4+6	87	81	80	38	34	22	0	0	0	"	"	2x6	"	"	½x4x31	16x20	14x14	8x10	12x10
	2x12	6+6	98	90	91	42	38	36	0	14	0	"	"	"	"	"	½x4x36	18x20	18x14	8x12	16x10
5/12 Slope	2x4	2x4	22	21	0	0	0	0	0	0	0	2x4	2x4	2x4	2x4	2x4	3/8x3½x15	8x12	8x8	8x8	8x8
	2x6	2x4	41	35	0	0	0	0	0	0	0	"	"	"	"	"	½x4x13	10x12	"	"	"
	2x6	2x6	46	45	44	0	18	0	0	0	0	"	"	"	"	"	½x4x16	10x16	10x8	"	"
	2x8	2x6	63	59	55	0	20	0	0	0	0	2x4	2x4	2x6	2x4	2x4	½x4x19	12x16	10x12	8x10	8x8
	2x10	4+4	86	82	69	37	25	0	0	0	0	"	"	2x8	4&4	"	½x4x26	14x20	14x14	8x12	10x8
	2x12	4+6	100+	100	100	47	42	30	0	0	0	"	"	4&4	"	"	½x4x31	18x20	16x10	8x10	12x10
	2x12	6+6	-	-	-	50	47	45	0	19	13	"	"	"	"	"	½x4x31	"	18x10	"	14x10

1400f Lumber

Slope	Top chord	Bottom chord	Truss spacing, ft. 2' — Ceiling dead load, psf 0	2' 5	2' 8	4' 0	4' 5	4' 8	8' 0	8' 5	8' 8	Web member sizes W1	W2	W3	W4	W5	Gusset Sizes, in. A T H W	B H W	C H W	D H W	E H W
			---Max. snow + roof dead load, psf---																		
3/12 Slope	2x4	2x4	0	20	18	0	0	0	0	0	0	2x4	2x4	2x4	2x4	2x4	3/8x3½x24	8x12	8x8	8x8	8x8
	2x6	2x4	0	29	17	0	0	0	0	0	0	"	"	"	"	"	½x4x17	10x12	"	"	"
	2x6	2x6	43	41	39	0	15	0	0	0	0	"	"	"	"	"	½x4x23	10x16	10x10	"	"
	2x8	2x6	49	45	41	0	16	0	0	0	0	2x4	2x4	2x4	2x4	2x4	½x4x25	12x16	10x10	8x8	8x8
	2x10	4+4	69	64	59	0	24	0	0	0	0	"	"	"	"	"	½x4x32	16x16	12x10	"	10x8
	2x12	4+6	87	80	77	38	34	29	0	0	0	"	"	"	"	"	½x4x39	16x20	14x12	8x10	12x8
	2x12	6+6	92	85	81	40	36	33	0	15	0	"	"	"	"	"	½x4x44	18x20	18x12	"	14x10
4/12 Slope	2x4	2x4	25	23	22	0	0	0	0	0	0	2x4	2x4	2x4	2x4	2x4	3/8x3½x22	8x12	8x8	8x8	8x8
	2x6	2x4	43	40	25	0	0	0	0	0	0	"	"	"	"	"	½x4x17	10x12	8x10	"	"
	2x6	2x6	51	49	47	0	20	14	0	0	0	"	"	"	"	"	½x4x21	12x12	10x10	"	8x10
	2x8	2x6	66	62	59	0	25	15	0	0	0	2x4	2x4	2x4	2x4	2x4	½x4x24	12x16	10x10	8x8	8x10
	2x10	4+4	91	84	81	39	32	15	0	0	0	"	"	2x6	"	"	½x4x33	14x20	12x14	8x12	12x10
	2x12	4+6	100+	100+	100+	50	46	39	0	15	0	"	"	"	"	"	2-½x6x28	18x20	16x14	"	14x10
	2x12	6+6	-	-	-	52	47	45	0	22	13	"	"	"	"	"	½x4x39	"	18x14	"	16x10
5/12 Slope	2x4	2x4	28	26	25	0	0	0	0	0	0	2x4	2x4	2x4	2x4	2x4	3/8x3½x19	8x12	8x8	8x8	8x8
	2x6	2x4	55	50	35	0	0	0	0	0	0	"	"	2x6	"	"	½x4x17	10x16	10x10	8x10	"
	2x6	2x6	57	55	53	25	22	21	0	0	0	"	"	"	4&4	"	½x4x19	"	10x12	"	"
	2x8	2x6	83	77	77	36	32	21	0	0	0	2x4	2x4	2x6	4&4	2x4	½x4x25	12x20	12x12	8x10	8x8
	2x10	4+4	100+	100+	100+	49	39	21	0	0	0	"	"	2x8	"	"	2-½x6x18	16x20	14x16	8x14	10x10
	2x12	4+6	-	-	-	61	57	49	0	19	0	"	"	4&4	"	"	½x4x36	18x20	16x12	8x10	12x10
	2x12	6+6	-	-	-	62	57	54	31	27	19	"	4&4	"	"	"	½x4x38	"	18x12	"	14x10

1600f Lumber

Slope	Top chord	Bottom chord	Truss spacing, ft. 2' — Ceiling dead load, psf 0	2' 5	2' 8	4' 0	4' 5	4' 8	8' 0	8' 5	8' 8	Web member sizes W1	W2	W3	W4	W5	Gusset Sizes, in. A T H W	B H W	C H W	D H W	E H W
			---Max. snow + roof dead load, psf---																		
3/12 Slope	2x4	2x4	27	25	23	0	0	0	0	0	0	2x4	2x4	2x4	2x4	2x4	½x3½x16	8x12	8x8	8x8	8x8
	2x6	2x4	39	36	29	0	0	0	0	0	0	"	"	"	"	"	½x4x20	10x12	8x10	"	"
	2x6	2x6	53	51	48	0	19	15	0	0	0	"	"	"	"	"	½x4x28	10x16	10x10	8x10	"
	2x8	2x6	58	54	51	0	20	15	0	0	0	2x4	2x4	2x4	2x4	2x4	½x4x29	12x16	10x10	8x10	8x8
	2x10	4+4	83	77	76	36	31	21	0	0	0	"	"	"	"	"	½x4x39	16x20	12x12	"	10x10
	2x12	4+6	100+	97	98	45	42	38	0	15	0	"	"	"	"	"	½x4x48	18x20	16x12	"	12x10
	2x12	6+6	-	100+	100+	49	45	43	0	20	14	"	"	"	"	"	½x4x53	20x20	18x14	"	14x12
4/12 Slope	2x4	2x4	30	28	27	0	0	0	0	0	0	2x4	2x4	2x4	2x4	2x4	½x3½x14	8x12	8x8	8x8	8x8
	2x6	2x4	52	49	44	0	17	0	0	0	0	"	"	"	"	"	½x4x20	12x12	10x10	"	"
	2x6	2x6	62	59	57	27	24	23	0	0	0	"	"	"	"	"	½x4x25	10x16	"	"	8x10
	2x8	2x6	80	76	72	34	31	29	0	13	0	2x4	2x4	2x4	2x4	2x4	½x4x29	12x20	10x10	8x8	8x10
	2x10	4+4	100+	100+	100+	48	45	41	0	19	13	"	"	"	"	"	2-½x6x22	16x20	14x12	8x10	12x10
	2x12	4+6	-	-	-	60	57	54	0	26	23	"	"	2x6	"	"	2-½x6x28	18x20	18x14	"	14x10
	2x12	6+6	-	-	-	63	58	56	31	27	25	"	"	2x6	4&4	"	2-½x6x34	18x22	20x14	8x12	16x12
5/12 Slope	2x4	2x4	33	32	31	0	0	0	0	0	0	2x4	2x4	2x4	4&4	2x4	3/8x3½x22	8x12	8x8	8x8	8x8
	2x6	2x4	66	60	51	0	0	0	0	0	0	"	"	"	"	"	½x4x20	10x16	10x10	"	"
	2x6	2x6	69	67	65	30	28	27	0	0	0	"	"	"	"	"	½x4x22	10x20	"	"	"
	2x8	2x6	100	93	94	43	39	30	0	0	0	2x4	2x4	2x6	4&4	2x4	2-½x6x15	14x16	12x12	8x12	8x10
	2x10	4+4	-	100+	100+	59	51	37	0	0	0	"	"	2x8	"	"	2-½x6x20	18x24	16x16	10x14	12x10
	2x12	4+6	-	-	-	74	68	62	37	28	14	"	"	"	"	"	2-½x6x22	20x24	18x16	10x14	14x10
	2x12	6+6				75	69	66	37	33	29	"	"	"	"	"	2-½x6x28	"	20x18	8x16	14x12

48' SPAN, 4-WEB

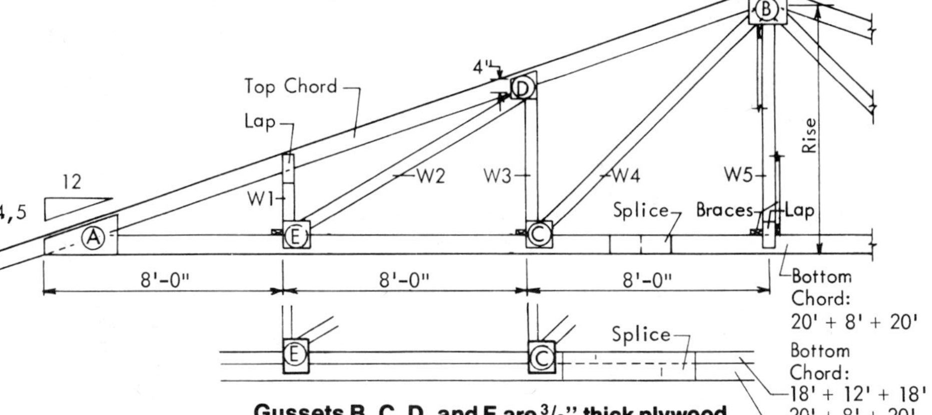

Gussets B, C, D, and E are 3/8" thick plywood.

4+4, 4+6, 6+6 indicates stacked lower chord.
4&4, 6&4, indicate double web; a 2x4 is attached to the web member to increase its stiffness.

Before selecting heel gusset A, see pages 8 and 9 .

Web Lengths

Roof Slope	Rise	Top Chord	W1	W2	W3	W4	W5
3/12	6'-0"	20'+5'	2'	9'	4'	10'+9'	6'
4/12	8'-0"	20'+6'	3'	10'+9'	5'	11'+10'	8'
5/12	10'-0"	14'+13'	3'	10'+9'	7'+6'	13'+12'	10'

1100f Lumber

See Ceiling Load Note, page 4.

Truss spacing, ft. — Ceiling dead load, psf — Max. snow + roof dead load, psf

Slope	Top chord	Bottom chord	2': 0	2': 5	2': 8	4': 0	4': 5	4': 8	8': 0	8': 5	8': 8	W1	W2	W3	W4	W5	A T H W	B H W	C H W	D H W	E H W
3/12 Slope	2x4	2x4	0	14	0	0	0	0	0	0	0	2x4	2x4	2x4	2x4	2x4	3/8x3½x19	8x12	8x8	8x8	8x8
	2x6	2x4	0	17	0	0	0	0	0	0	0	"	"	"	"	"	3/8x4x22	10x12	"	"	"
	2x6	2x6	33	29	26	0	0	0	0	0	0	"	"	"	"	"	3/8x4x32	"	8x10	"	"
	2x8	2x6	36	31	27	0	0	0	0	0	0	2x4	2x4	2x4	2x4	2x4	3/8x4x34	12x12	8x10	8x8	8x8
	2x10	4+4	50	47	41	0	14	0	0	0	0	"	"	"	"	"	½x4x28	14x16	12x10	"	10x8
	2x12	4+6	64	59	56	0	23	13	0	0	0	"	"	"	"	"	½x4x36	16x16	14x10	"	12x8
	2x12	6+6	73	66	63	31	27	23	0	0	0	"	"	"	4&4	"	½x4x44	18x16	16x12	8x10	14x10
4/12 Slope	2x4	2x4	0	17	0	0	0	0	0	0	0	2x4	2x4	2x4	2x4	2x4	3/8x3½x17	8x12	8x8	8x8	8x8
	2x6	2x4	0	25	0	0	0	0	0	0	0	"	"	"	"	"	3/8x4x22	10x12	"	"	"
	2x6	2x6	39	37	35	0	13	0	0	0	0	"	"	"	"	"	½x4x18	"	8x10	"	"
	2x8	2x6	48	45	42	0	13	0	0	0	0	2x4	2x4	2x4	2x4	2x4	½x4x19	14x12	10x10	8x8	8x8
	2x10	4+4	66	61	52	0	17	0	0	0	0	"	"	2x6	"	"	½x4x25	14x16	12x12	8x10	10x10
	2x12	4+6	84	77	76	36	31	20	0	0	0	"	"	"	4&4	"	½x4x31	16x20	14x14	"	12x10
	2x12	6+6	94	86	83	41	36	32	0	12	0	"	"	"	4&4	"	½x4x36	18x16	18x14	8x12	16x10
5/12 Slope	2x4	2x4	20	19	0	0	0	0	0	0	0	2x4	2x4	2x4	4&4	2x4	3/8x3½x15	8x12	8x8	8x8	8x8
	2x6	2x4	40	32	0	0	0	0	0	0	0	"	"	"	"	"	½x4x13	10x12	"	"	"
	2x6	2x6	43	42	41	0	17	0	0	0	0	"	"	2x6	"	"	½x4x16	10x16	10x10	8x10	"
	2x8	2x6	61	57	52	0	18	0	0	0	0	2x4	2x4	2x6	4&4	2x4	½x4x19	12x16	10x12	8x10	8x8
	2x10	4+4	83	78	64	36	21	0	0	0	0	"	"	2x8	"	"	½x4x26	14x20	14x14	8x12	10x8
	2x12	4+6	100+	95	96	45	39	26	0	0	0	"	"	4&4	"	"	½x4x31	18x20	16x10	8x10	12x10
	2x12	6+6	–	100+	100+	48	44	42	0	16	0	"	"	"	"	"	½x4x35	"	18x12	"	14x10

1400f Lumber

Slope	Top chord	Bottom chord	2' Ceiling dead load 0 psf	2' 5	2' 8	4' 0	4' 5	4' 8	8' 0	8' 5	8' 8	W1	W2	W3	W4	W5	A T H W	B H W	C H W	D H W	E H W
			---Max. snow + roof dead load, psf---									Web member sizes					Gusset Sizes, in.				
3/12 Slope	2x4	2x4	0	17	12	0	0	0	0	0	0	2x4	2x4	2x4	2x4	2x4	3/8x3½x22	8x12	8x8	8x8	8x8
	2x6	2x4	0	28	12	0	0	0	0	0	0	"	"	"	"	"	½x4x17	10x12	"	"	"
	2x6	2x6	41	39	36	0	14	0	0	0	0	"	"	"	"	"	½x4x23	10x16	10x10	"	"
	2x8	2x6	47	43	39	0	15	0	0	0	0	2x4	2x4	2x4	2x4	2x4	½x4x25	12x16	10x10	8x8	8x8
	2x10	4+4	66	61	56	0	22	0	0	0	0	"	"	"	"	"	½x4x32	16x16	12x10	"	10x8
	2x12	4+6	83	77	73	36	32	25	0	0	0	"	"	"	"	"	½x4x40	18x20	14x12	8x10	12x10
	2x12	6+6	88	81	77	38	34	32	0	15	0	"	"	"	"	"	½x4x44	"	18x12	"	14x10
4/12 Slope	2x4	2x4	23	22	21	0	0	0	0	0	0	2x4	2x4	2x4	2x4	2x4	3/8x3½x19	8x12	8x8	8x8	8x8
	2x6	2x4	42	38	19	0	0	0	0	0	0	"	"	"	"	"	½x4x17	10x16	10x10	8x10	"
	2x6	2x6	48	46	44	0	18	12	0	0	0	"	"	"	"	"	½x4x19	"	10x12	"	"
	2x8	2x6	63	59	56	0	23	12	0	0	0	2x4	2x4	2x4	2x4	2x4	½x4x25	12x20	12x14	8x12	8x8
	2x10	4+4	87	80	76	38	29	0	0	0	0	"	"	2x6	"	"	2–½x6x18	16x20	14x12	8x10	10x10
	2x12	4+6	100+	100+	100+	48	44	35	0	13	0	"	"	"	"	"	½x4x36	18x20	16x12	"	12x10
	2x12	6+6	–	–	–	49	45	43	0	21	0	"	"	"	"	"	½x4x38	20x24	20x12	"	14x10
5/12 Slope	2x4	2x4	26	25	23	0	0	0	0	0	0	2x4	2x4	2x4	4&4	2x4	3/8x3½x21	8x12	8x8	8x8	8x8
	2x6	2x4	53	47	26	0	0	0	0	0	0	"	"	2x6	"	"	½x4x17	10x12	8x10	"	"
	2x6	2x6	53	51	50	23	21	17	0	0	0	"	"	"	"	"	½x4x21	12x12	10x10	"	8x10
	2x8	2x6	80	74	73	34	29	17	0	0	0	2x4	2x4	2x8	4&4	2x4	½x4x25	12x16	10x10	8x8	8x10
	2x10	4+4	100+	100+	97	47	36	15	0	0	0	"	"	4&4	"	"	½x4x33	14x20	12x14	8x12	12x10
	2x12	4+6	–	–	100+	59	54	45	0	17	0	"	"	"	"	"	2–½x6x28	18x20	16x14	"	14x10
	2x12	6+6	–	–	–	59	54	52	29	25	15	"	4&4	"	"	"	½x4x39	"	18x14	"	16x10

1600f Lumber

Slope	Top chord	Bottom chord	2' Ceiling dead load 0 psf	2' 5	2' 8	4' 0	4' 5	4' 8	8' 0	8' 5	8' 8	W1	W2	W3	W4	W5	A T H W	B H W	C H W	D H W	E H W
			---Max. snow + roof dead load, psf---									Web member sizes					Gusset Sizes, in.				
3/12 Slope	2x4	2x4	25	23	21	0	0	0	0	0	0	2x4	2x4	2x4	2x4	2x4	½x3½x16	8x12	8x8	8x8	8x8
	2x6	2x4	37	34	23	0	0	0	0	0	0	"	"	"	"	"	½x4x20	10x16	8x10	"	"
	2x6	2x6	50	48	46	0	18	13	0	0	0	"	"	"	"	"	½x4x28	"	10x10	8x10	"
	2x8	2x6	56	51	48	0	19	13	0	0	0	2x4	2x4	2x4	2x4	2x4	½x4x29	12x16	10x10	8x10	8x8
	2x10	4+4	79	74	73	34	28	14	0	0	0	"	"	"	"	"	½x4x39	16x20	14x12	"	10x10
	2x12	4+6	100	92	93	43	40	35	0	14	0	"	"	"	"	"	½x4x48	18x20	16x12	"	12x10
	2x12	6+6	–	100	100+	47	43	40	0	19	12	"	"	"	"	"	½x4x54	20x20	18x14	10x10	14x12
4/12 Slope	2x4	2x4	28	26	25	0	0	0	0	0	0	2x4	2x4	2x4	4&4	2x4	½x3½x14	8x12	8x8	8x8	8x8
	2x6	2x4	50	47	42	0	16	0	0	0	0	"	"	"	"	"	½x4x20	12x12	10x10	"	"
	2x6	2x6	58	55	53	25	22	21	0	0	0	"	"	"	"	"	½x4x25	10x16	"	"	8x10
	2x8	2x6	77	73	69	33	30	27	0	12	0	2x4	2x4	2x4	4&4	2x4	½x4x30	12x20	10x10	8x10	8x10
	2x10	4+4	100+	99	100	46	43	39	0	18	0	"	"	2x6	"	"	2–½x6x22	16x20	14x14	8x12	12x10
	2x12	4+6	–	–	–	58	54	51	0	25	22	"	"	"	"	"	2–½x6x28	18x20	18x16	"	14x10
	2x12	6+6	–	–	–	61	56	53	30	26	24	"	"	"	"	"	2–½x6x34	18x22	20x16	8x14	16x12
5/12 Slope	2x4	2x4	31	30	29	0	0	0	0	0	0	2x4	2x4	2x4	4&4	2x4	3/8x3½x22	8x12	8x8	8x8	8x8
	2x6	2x4	63	57	46	0	0	0	0	0	0	"	"	2x6	"	"	½x4x20	10x16	10x10	8x10	"
	2x6	2x6	65	62	61	28	26	25	0	0	0	"	"	"	"	"	½x4x22	10x20	"	"	"
	2x8	2x6	96	89	90	41	38	26	0	0	0	2x4	2x4	2x6	4&4	2x4	2–½x6x15	14x20	12x12	8x12	8x10
	2x10	4+4	100+	100+	100+	57	48	31	0	0	0	"	"	2x8	"	"	2–½x6x21	18x24	16x16	10x14	12x10
	2x12	4+6	–	–	–	71	65	55	35	25	0	"	4&4	4&4	"	"	2–½x6x22	20x24	18x12	10x10	14x10
	2x12	6+6	–	–	–	71	66	63	35	32	24	"	"	"	"	"	2–½x6x28	"	20x14	8x12	14x12

50' SPAN, 4-WEB

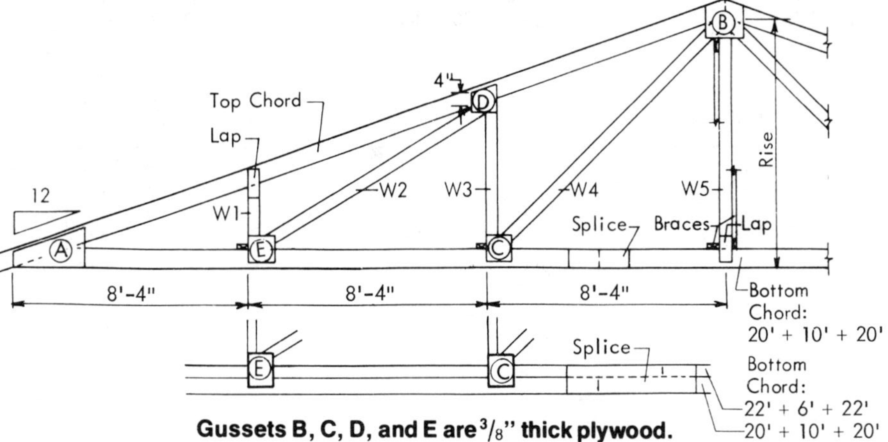

Gussets B, C, D, and E are 3/8" thick plywood.

4+4, 4+6, 6+6 indicates stacked lower chord.
4&4, 6&4, indicate double web; a 2x4 is attached to the web member to increase its stiffness.

Before selecting heel gusset A, see pages 8 and 9 .

Web Lengths

Roof Slope	Rise	Top Chord	W1	W2	W3	W4	W5
3/12	6'-3"	20'+6'	2'	9'	4'	10'	6'
4/12	8'-4"	14'+13'	3'	10'+9'	6'	12'+11'	8'
5/12	10'-5"	14'+14'	4'	11'+10'	7'+6'	13'+12'	11'

1100f Lumber

			Truss spacing, ft.									Web member sizes					Gusset Sizes, in.				
			2'			4'			8'												
			Ceiling dead load, psf														A	B	C	D	E
	Top chord	Bottom chord	0	5	8	0	5	8	0	5	8	W1	W2	W3	W4	W5	T H W	H W	H W	H W	H W
			---Max. snow + roof dead load, psf---																		
3/12 Slope	2x4	2x4	0	0	0	0	0	0	0	0	0	2x4	2x4	2x4	2x4	2x4	3/8x3½x16	8x12	8x8	8x8	8x8
	2x6	2x4	0	13	0	0	0	0	0	0	0	"	"	"	"	"	3/8x4x22	10x12	"	"	"
	2x6	2x6	0	28	24	0	0	0	0	0	0	"	"	"	"	"	3/8x4x32	"	8x10	"	"
	2x8	2x6	0	29	26	0	0	0	0	0	0	2x4	2x4	2x4	2x4	2x4	3/8x4x34	12x12	8x10	8x8	8x8
	2x10	4+4	48	45	37	0	0	0	0	0	0	"	"	"	"	"	½x4x28	14x16	12x10	"	10x8
	2x12	4+6	61	56	54	0	21	0	0	0	0	"	"	"	"	"	½x4x36	16x16	14x10	"	12x8
	2x12	6+6	70	64	60	0	26	20	0	0	0	"	"	"	"	"	½x4x44	18x16	16x12	8x10	14x10
4/12 Slope	2x4	2x4	0	16	0	0	0	0	0	0	0	2x4	2x4	2x4	2x4	2x4	3/8x3½x17	8x12	8x8	8x8	8x8
	2x6	2x4	0	20	0	0	0	0	0	0	0	"	"	"	"	"	3/8x4x22	10x12	"	"	"
	2x6	2x6	37	35	33	0	0	0	0	0	0	"	"	"	"	"	½x4x18	"	8x10	"	"
	2x8	2x6	47	43	40	0	12	0	0	0	0	2x4	2x4	2x4	2x4	2x4	½x4x19	14x12	10x10	8x8	8x8
	2x10	4+4	64	58	49	0	13	0	0	0	0	"	"	2x6	"	"	½x4x25	14x16	12x12	8x10	10x10
	2x12	4+6	80	74	73	35	28	16	0	0	0	"	"	"	4&4	"	½x4x31	16x20	14x14	"	12x10
	2x12	6+6	90	83	79	39	35	30	0	0	0	"	"	"	"	"	½x4x36	18x20	18x14	8x12	16x10
5/12 Slope	2x4	2x4	19	18	0	0	0	0	0	0	0	2x4	2x4	2x4	2x4	2x4	3/8x3½x15	8x12	8x8	8x8	8x8
	2x6	2x4	38	27	0	0	0	0	0	0	0	"	"	2x6	"	"	½x4x13	10x12	8x10	8x10	"
	2x6	2x6	41	39	38	0	16	0	0	0	0	"	"	"	"	"	½x4x16	10x16	10x10	"	"
	2x8	2x6	59	54	50	0	15	0	0	0	0	2x4	2x4	2x8	2x4	2x4	½x4x19	12x16	10x14	8x12	8x8
	2x10	4+4	80	74	59	34	19	0	0	0	0	"	"	4&4	"	"	½x4x26	14x20	14x10	8x10	10x10
	2x12	4+6	100	91	91	43	36	21	0	0	0	"	"	"	"	"	½x4x32	18x20	16x10	"	12x10
	2x12	6+6	-	98	99	45	42	38	0	14	0	"	"	"	"	"	½x4x35	"	18x12	"	14x10

1400f Lumber

Slope	Top chord	Bottom chord	Truss spacing, ft. 2' Ceiling dead load, psf 0	2' 5	2' 8	4' 0	4' 5	4' 8	8' 0	8' 5	8' 8	Web member sizes W1	W2	W3	W4	W5	Gusset Sizes, in. A T H W	B H W	C H W	D H W	E H W
			---Max. snow + roof dead load, psf---																		
3/12 Slope	2x4	2x4	0	17	0	0	0	0	0	0	0	2x4	2x4	2x4	2x4	2x4	3/8x3½x20	8x12	8x8	8x8	8x8
	2x6	2x4	0	26	0	0	0	0	0	0	0	"	"	"	"	"	½x4x17	10x12	"	"	"
	2x6	2x6	38	36	34	0	13	0	0	0	0	"	"	"	"	"	½x4x23	10x16	10x10	"	"
	2x8	2x6	45	40	37	0	14	0	0	0	0	2x4	2x4	2x4	2x4	2x4	½x4x25	12x16	10x10	8x8	8x8
	2x10	4+4	63	59	53	0	19	0	0	0	0	"	"	"	"	"	½x4x33	16x16	12x12	8x10	10x8
	2x12	4+6	80	74	70	0	31	22	0	0	0	"	"	"	"	"	½x4x40	18x20	14x12	"	12x10
	2x12	6+6	85	78	74	37	32	30	0	14	0	"	4&4	"	"	"	½x4x44	"	18x12	"	14x10
4/12 Slope	2x2	2x4	0	20	14	0	0	0	0	0	0	2x4	2x4	2x4	2x4	2x4	3/8x3½x20	8x12	8x8	8x8	8x8
	2x6	2x4	0	36	13	0	0	0	0	0	0	"	"	"	"	"	½x4x17	10x12	8x10	"	"
	2x6	2x6	45	43	42	0	17	0	0	0	0	"	"	"	"	"	½x4x21	12x12	10x10	"	8x10
	2x8	2x6	61	57	54	0	21	0	0	0	0	2x4	2x4	2x4	2x4	2x4	½x4x25	12x16	10x10	8x8	8x10
	2x10	4+4	84	80	71	0	27	17	0	0	0	"	"	2x6	"	"	½x4x34	16x16	14x14	8x12	12x10
	2x12	4+6	100+	97	98	46	42	32	0	0	0	"	"	"	4&4	"	2-½x6x28	18x20	16x14	"	14x10
	2x12	6+6	-	100+	-	47	44	41	0	20	0	"	"	"	"	"	½x4x40	"	18x14	"	16x10
5/12 Slope	2x4	2x4	24	22	21	0	0	0	0	0	0	2x4	2x4	2x4	4&4	2x4	3/8x3½x18	8x12	8x8	8x8	8x8
	2x6	2x4	51	45	19	0	0	0	0	0	0	"	"	2x6	"	"	½x4x26	10x16	10x10	8x10	"
	2x6	2x6	50	48	47	0	20	14	0	0	0	"	"	"	"	"	½x4x19	"	10x12	"	"
	2x8	2x6	77	71	70	0	27	14	0	0	0	2x4	2x4	2x8	4&4	2x4	½x4x25	12x20	12x14	8x12	8x8
	2x10	4+4	100+	100	90	45	33	0	0	0	0	"	"	4&4	"	"	2-½x6x18	16x20	14x12	8x10	10x10
	2x12	4+6	-	-	100+	57	52	40	0	15	0	"	"	"	"	"	½x4x36	18x20	16x12	"	12x10
	2x12	6+6	-	-	-	56	52	50	0	24	12	"	"	"	"	"	½x4x38	20x24	20x12	10x10	16x10

1600f Lumber

Slope	Top chord	Bottom chord	Truss spacing, ft. 2' Ceiling dead load, psf 0	2' 5	2' 8	4' 0	4' 5	4' 8	8' 0	8' 5	8' 8	Web member sizes W1	W2	W3	W4	W5	Gusset Sizes, in. A T H W	B H W	C H W	D H W	E H W
			---Max. snow + roof dead load, psf---																		
3/12 Slope	2x4	2x4	0	21	19	0	0	0	0	0	0	2x4	2x4	2x4	2x4	2x4	3/8x3½x27	8x12	8x8	8x8	8x8
	2x6	2x4	0	33	18	0	0	0	0	0	0	"	"	"	"	"	½x4x20	10x16	8x10	"	"
	2x6	2x6	47	45	44	0	17	0	0	0	0	"	"	"	"	"	½x4x28	"	10x10	8x10	"
	2x8	2x6	54	49	46	0	18	0	0	0	0	2x4	2x4	2x4	2x4	2x4	½x4x29	12x16	10x10	8x10	8x10
	2x10	4+4	76	71	69	0	26	0	0	0	0	"	"	"	"	"	½x4x39	16x20	14x12	"	10x10
	2x12	4+6	96	89	89	42	38	32	0	12	0	"	"	"	"	"	½x4x48	18x20	16x12	"	12x10
	2x12	6+6	100+	96	97	45	41	38	0	18	0	"	"	"	"	"	½x4x54	20x20	18x14	10x10	16x10
4/12 Slope	2x4	2x4	26	25	24	0	0	0	0	0	0	2x4	2x4	2x4	4&4	2x4	½x3½x14	8x12	8x8	8x8	8x8
	2x6	2x4	48	45	40	0	14	0	0	0	0	"	"	"	"	"	½x4x21	12x12	10x10	"	8x10
	2x6	2x6	54	52	50	0	21	20	0	0	0	"	"	"	"	"	½x4x24	10x16	"	"	"
	2x8	2x6	74	70	66	0	29	25	0	0	0	2x4	2x4	2x4	4&4	2x4	½x4x30	12x20	10x10	8x10	8x10
	2x10	4+4	-	94	95	44	40	36	0	17	0	"	"	2x6	"	"	2-½x6x22	16x20	14x14	8x12	12x10
	2x12	4+6	-	-	-	55	52	49	0	24	20	"	"	"	"	"	2-½x6x28	18x22	18x16	8x14	14x12
	2x12	6+6	-	-	-	58	54	51	0	25	22	"	"	"	"	"	2-½x6x34	"	20x16	8x16	16x12
5/12 Slope	2x4	2x4	29	28	27	0	0	0	0	0	0	2x4	2x4	2x4	4&4	2x4	3/8x3½x21	8x12	8x8	8x8	8x8
	2x6	2x4	61	55	36	0	0	0	0	0	0	"	"	2x6	"	"	½x4x20	10x16	10x12	8x10	"
	2x6	2x6	61	58	57	26	24	23	0	0	0	"	"	"	"	"	½x4x22	"	"	"	"
	2x8	2x6	92	86	86	40	36	22	0	0	0	2x4	2x4	2x8	4&4	2x4	2-½x6x15	14x20	12x14	8x14	8x10
	2x10	4+4	100+	100+	100+	55	44	22	0	0	0	"	"	4&4	"	"	2-½x6x21	18x24	16x12	8x12	10x12
	2x12	4+6	-	-	-	68	62	52	34	22	0	"	4&4	"	"	"	2-½x6x22	20x24	18x12	"	12x12
	2x12	6+6	-	-	-	68	63	60	34	30	21	"	"	"	"	"	2-½x6x28	"	20x14	"	14x12

46' SPAN, 6-WEB

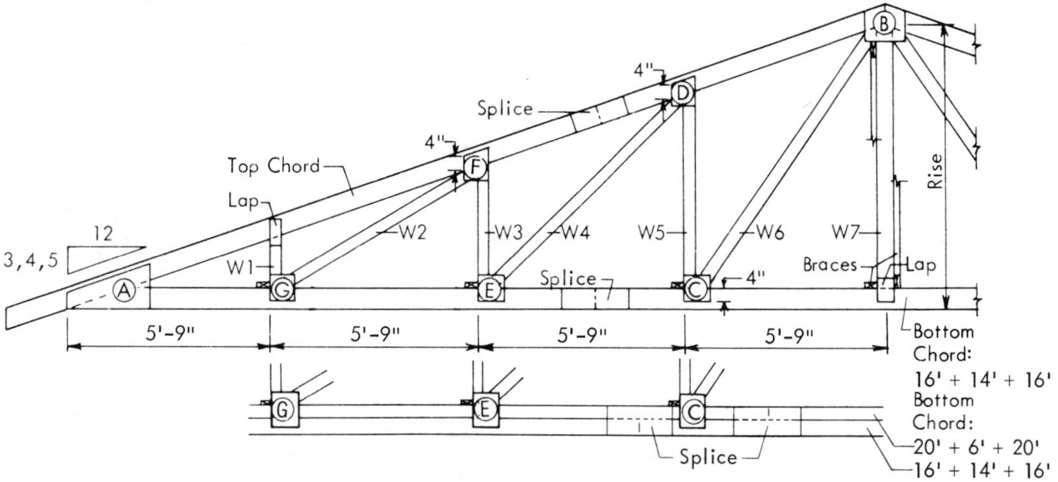

Top Chord
Lap
W1 W2 W3 W4 W5 W6 W7
Splice
4"
Braces
Rise
12
3, 4, 5
A B C D E F G
5'-9" 5'-9" 5'-9" 5'-9"
Bottom Chord: 16' + 14' + 16'
Bottom Chord: 20' + 6' + 20'
16' + 14' + 16'
Splice

4+4, 4+6, 6+6 indicates stacked lower chord.
4&4, 6&4, indicate double web; a 2x4 is attached to the web member to increase its stiffness.

Before selecting heel gusset A, see pages 8 and 9 .

2 web lengths indicate double member, see page 10.

Web Lengths

Gussets B, C, D, E, F, and G are 3/8" thick plywood.

Roof Slope	Rise	Top Chord	W1	W2	W3	W4	W5	W6	W7
3/12	5'-9"	16'+8"	2'	6'	3'	7'	4'	8'	6'
4/12	7'-8"	16'+9'	2'	7'	4'	8'	6'	10'+9'	8'
5/12	9'-7"	18'+8'	3'	7'	5'	9'	7'+6'	11'+10'	10'

1100f Lumber

			Truss spacing, ft.									Web member sizes							Gusset Sizes, in.						
			2'			4'			8'																
			Ceiling dead load, psf																A	B	C	D	E	F	G
Slope	Top chord	Bottom chord	0	5	8	0	5	8	0	5	8	W1	W2	W3	W4	W5	W6	W7	T H W	H W	H W	H W	H W	H W	H W
			—Max. snow + roof dead load, psf—																						
3/12 Slope	2x4	2x4	22	18	15	0	0	0	0	0	0	2x4	2x4	2x4	2x4	2x4	2x4	2x4	3/8x3½x22	8x12	8x8	8x8	8x8	8x8	8x8
	2x6	2x4	23	20	17	0	0	0	0	0	0	"	"	"	"	"	"	"	3/8x4x22	10x12	"	"	"	"	"
	2x6	2x6	33	29	27	14	0	0	0	0	0	"	"	"	"	"	"	"	3/8x4x32	"	"	"	"	"	"
	2x8	2x6	35	31	29	15	0	0	0	0	0	2x4	2x4	2x4	2x4	2x4	2x4	2x4	3/8x4x33	12x12	8x8	8x8	8x8	8x8	8x8
	2x10	4+4	51	47	45	22	19	14	0	0	0	"	"	"	"	"	"	"	½x4x28	14x16	12x10	"	10x8	"	10x8
	2x12	6+4	66	61	58	28	25	22	0	0	0	"	"	"	"	"	"	"	½x4x36	16x16	14x10	"	12x10	"	12x8
	2x12	6+6	74	69	66	32	28	26	0	12	0	"	"	"	"	"	"	"	½x4x44	16x20	16x10	8x10	16x10	"	14x8
4/12 Slope	2x4	2x4	26	23	22	0	0	0	0	0	0	2x4	2x4	2x4	2x4	2x4	2x4	2x4	3/8x3½x21	8x12	8x8	8x8	8x8	8x8	8x8
	2x6	2x4	31	27	23	0	0	0	0	0	0	"	"	"	"	"	"	"	3/8x4x22	10x12	"	"	"	"	"
	2x6	2x6	45	42	39	19	15	13	0	0	0	"	"	"	"	"	"	"	½x4x18	"	10x8	"	"	"	"
	2x8	2x6	48	44	43	21	17	15	0	0	0	2x4	2x4	2x4	2x4	2x4	2x4	2x4	½x4x19	12x12	10x8	8x8	8x8	8x8	8x8
	2x10	4+4	66	62	60	29	26	20	0	0	0	"	"	"	"	2x6	"	"	½x4x25	14x16	12x12	8x10	10x10	"	10x8
	2x12	6+4	85	79	76	37	33	31	18	14	0	"	"	"	"	"	4&4	"	½x4x30	16x16	14x12	8x12	14x10	"	12x8
	2x12	6+6	97	90	87	42	39	36	21	17	14	"	"	"	"	"	"	"	½x4x36	18x16	18x4	"	16x10	"	14x8
5/12 Slope	2x4	2x4	29	27	25	13	0	0	0	0	0	2x4	2x4	2x4	2x4	2x4	4&4	2x4	3/8x3½x19	8x12	8x8	8x8	8x8	8x8	8x8
	2x6	2x4	38	33	29	16	0	0	0	0	0	"	"	"	"	2x6	"	"	3/8x4x21	10x12	8x10	8x10	"	"	"
	2x6	2x6	57	53	51	25	21	18	0	0	0	"	"	"	"	"	"	"	½x4x19	10x16	10x10	8x12	"	"	"
	2x8	2x6	59	56	54	26	22	20	0	0	0	2x4	2x4	2x4	2x4	2x8	4&4	2x4	½x4x18	12x16	10x12	8x14	8x8	8x8	8x8
	2x10	4+4	82	78	75	35	33	25	17	0	0	"	"	"	4&4	4&4	"	"	½x4x25	14x16	12x10	"	12x8	"	10x8
	2x12	6+4	100+	98	95	45	42	41	22	19	0	"	"	"	"	"	"	"	½x4x30	16x20	14x10	8x10	14x10	"	12x8
	2x12	6+6	–	–	100+	52	48	46	26	22	19	"	"	2x6	"	"	"	"	½x4x36	18x16	18x10	"	18x12	8x10	14x8

1400f Lumber

	Top chord	Bottom chord	Truss spacing 2', ceiling dead load 0 psf	2', 5	2', 8	4', 0	4', 5	4', 8	8', 0	8', 5	8', 8	W1	W2	W3	W4	W5	W6	W7	A T H W	B H W	C H W	D H W	E H W	F H W	G H W
			—Max. snow + roof dead load, psf—									Web member sizes							Gusset Sizes, in.						
3/12 Slope	2x4	2x4	27	24	22	0	0	0	0	0	0	2x4	2x4	2x4	2x4	2x4	2x4	2x4	3/8x3½x28	8x12	8x8	8x8	8x8	8x8	8x8
	2x6	2x4	30	27	25	0	0	0	0	0	0	"	"	"	"	"	"	"	½x4x16	10x12	"	"	"	"	"
	2x6	2x6	43	40	37	19	15	12	0	0	0	"	"	"	"	"	"	"	½x4x23	10x16	8x10	"	"	"	"
	2x8	2x6	46	42	40	20	16	13	0	0	0	2x4	2x4	2x4	2x4	2x4	2x4	2x4	½x4x24	12x16	8x10	8x8	8x8	8x8	8x8
	2x10	4+4	67	62	60	29	25	23	0	0	0	"	"	"	"	"	"	"	½x4x31	14x16	12x10	8x10	10x10	"	10x8
	2x12	4+6	87	80	77	37	34	31	0	15	0	"	"	"	"	"	"	"	½x4x39	16x20	14x10	"	14x10	"	12x8
	2x12	6+6	93	85	85	40	36	33	20	15	12	"	"	"	"	"	4&4	"	½x4x44	18x16	16x12	"	16x10	"	14x8
4/12 Slope	2x4	2x4	32	29	27	14	0	0	0	0	0	2x4	2x4	2x4	2x4	2x4	2x4	2x4	½x3½x14	8x12	8x8	8x8	8x8	8x8	8x8
	2x6	2x4	40	38	34	17	13	0	0	0	0	"	"	"	"	"	"	"	½x4x16	10x12	"	"	"	"	"
	2x6	2x6	59	55	53	25	22	19	12	0	0	"	"	"	"	"	"	"	½x4x23	10x16	10x10	"	8x10	"	"
	2x8	2x6	63	58	56	27	24	21	0	0	0	2x4	2x4	2x4	2x4	2x4	2x4	2x4	½x4x23	12x16	10x10	8x8	8x10	8x8	8x8
	2x10	4+4	87	82	79	38	35	33	0	13	0	"	"	"	"	2x6	"	"	½x4x31	14x16	12x12	8x12	12x10	"	10x8
	2x12	4+6	100+	100+	100+	48	45	43	24	21	15	"	"	"	"	"	4&4	"	½x4x40	18x16	16x14	8x14	14x10	"	12x8
	2x12	6+6	–	–	–	52	48	46	26	22	19	"	"	"	"	"	"	"	½x4x40	"	18x14	"	16x10	"	14x8
5/12 Slope	2x4	2x4	36	33	31	15	12	0	0	0	0	2x4	2x4	2x4	2x4	2x4	2x4	2x4	½x3½x13	8x12	8x8	8x8	8x8	8x8	8x8
	2x6	2x4	50	46	42	22	16	0	0	0	0	"	"	"	"	2x6	"	"	½x4x16	10x12	8x10	8x10	"	"	"
	2x6	2x6	70	65	63	30	27	25	0	0	0	"	"	"	"	2x8	"	"	½x4x22	10x16	10x12	8x12	10x8	"	"
	2x8	2x6	78	73	70	34	30	29	0	13	0	2x4	2x4	2x4	2x4	2x8	2x4	2x4	½x4x23	12x16	10x12	8x12	10x8	8x8	8x8
	2x10	4+4	100+	100+	99	47	44	42	23	17	12	"	"	"	"	4&4	4&4	"	½x4x32	16x16	14x10	8x10	12x10	"	10x8
	2x12	4+6	–	–	–	59	56	54	29	26	20	"	"	2x6	"	"	"	"	2–½x6x22	18x20	18x12	8x12	16x12	8x10	12x8
	2x12	6+6	–	–	–	64	58	56	32	27	25	"	"	"	"	"	"	"	½x4x39	"	20x12	"	18x12	"	14x8

1600f Lumber

	Top chord	Bottom chord	Truss spacing 2', ceiling dead load 0 psf	2', 5	2', 8	4', 0	4', 5	4', 8	8', 0	8', 5	8', 8	W1	W2	W3	W4	W5	W6	W7	A T H W	B H W	C H W	D H W	E H W	F H W	G H W
			—Max. snow + roof dead load, psf—									Web member sizes							Gusset Sizes, in.						
3/12 Slope	2x4	2x4	33	30	28	0	0	0	0	0	0	2x4	2x4	2x4	2x4	2x4	2x4	2x4	½x3½x19	8x12	8x8	8x8	8x8	8x8	8x8
	2x6	2x4	36	33	31	0	12	0	0	0	0	"	"	"	"	"	"	"	½x4x19	10x12	8x10	"	"	"	"
	2x6	2x6	52	48	46	22	19	16	0	0	0	"	"	"	"	"	"	"	½x4x27	10x16	10x10	"	8x10	"	8x8
	2x8	2x6	55	51	49	24	20	17	0	0	0	2x4	2x4	2x4	2x4	2x4	2x4	2x4	½x4x28	12x16	10x10	8x8	8x10	8x8	8x8
	2x10	4+4	80	75	72	35	31	29	0	13	0	"	"	"	"	"	"	"	½x4x37	14x20	12x10	8x10	12x10	"	10x8
	2x12	4+6	100+	96	97	45	41	39	22	19	16	"	"	"	"	"	"	"	½x4x47	18x20	16x12	10x10	14x10	"	12x8
	2x12	6+6	–	100+	100+	49	45	42	24	20	17	"	"	"	"	"	"	"	½x4x53	"	18x12	"	16x12	"	14x8
4/12 Slope	2x4	2x4	39	37	34	17	14	0	0	0	0	2x4	2x4	2x4	2x4	2x4	2x4	2x4	½x3½x17	8x12	10x8	8x8	8x8	8x8	8x8
	2x6	2x4	48	46	43	21	16	0	0	0	0	"	"	"	"	"	"	"	½x4x19	10x12	10x8	"	"	"	"
	2x6	2x6	71	66	63	31	27	24	15	0	0	"	"	"	"	"	4&4	"	¼x4x27	10x16	10x10	"	8x10	"	"
	2x8	2x6	75	70	67	32	29	27	16	12	0	2x4	2x4	2x4	2x4	2x4	4&4	2x4	½x4x27	12x16	10x10	8x8	8x10	8x8	8x8
	2x10	4+4	100+	98	95	45	43	41	22	19	0	"	"	"	"	2x6	"	"	2–½x6x22	16x16	14x14	8x14	12x10	"	10x8
	2x12	4+6	–	100+	100+	58	54	52	29	25	23	"	"	"	"	"	"	"	2–½x6x28	18x20	18x14	10x12	16x12	"	12x8
	2x12	6+6	–	–	–	65	59	56	32	28	25	"	"	"	"	"	"	"	2–½x6x34	"	20x14	"	8x12	"	14x8
5/12 Slope	2x4	2x4	44	42	40	19	16	0	0	0	0	2x4	2x4	2x4	2x4	2x4	2x4	2x4	½x3½x16	8x12	8x8	8x8	8x8	8x8	8x8
	2x6	2x4	60	55	51	26	20	0	0	0	0	"	"	"	"	2x6	"	"	½x4x18	10x16	8x12	8x10	"	"	"
	2x6	2x6	86	81	78	37	34	32	18	15	0	"	"	"	"	"	4&4	"	2–½x6x14	10x20	10x12	8x10	10x10	"	"
	2x8	2x6	93	87	84	40	38	35	20	17	0	2x4	2x4	2x4	2x4	2x8	4&4	2x4	2–½x6x14	12x20	10x12	8x12	10x10	8x8	8x8
	2x10	4+4	100+	100+	100+	56	53	52	28	24	12	"	"	"	"	4&4	"	"	2–½x6x19	16x20	14x10	8x12	14x10	"	10x8
	2x12	4+6	–	–	–	71	67	65	35	32	29	"	"	"	"	"	"	"	2–½x6x22	18x20	18x12	10x10	16x10	"	12x8
	2x12	6+6	–	–	–	78	72	69	39	35	32	"	"	2x6	"	"	"	"	2–½x6x28	"	20x12	"	18x14	8x10	14x8

48' SPAN, 6-WEB

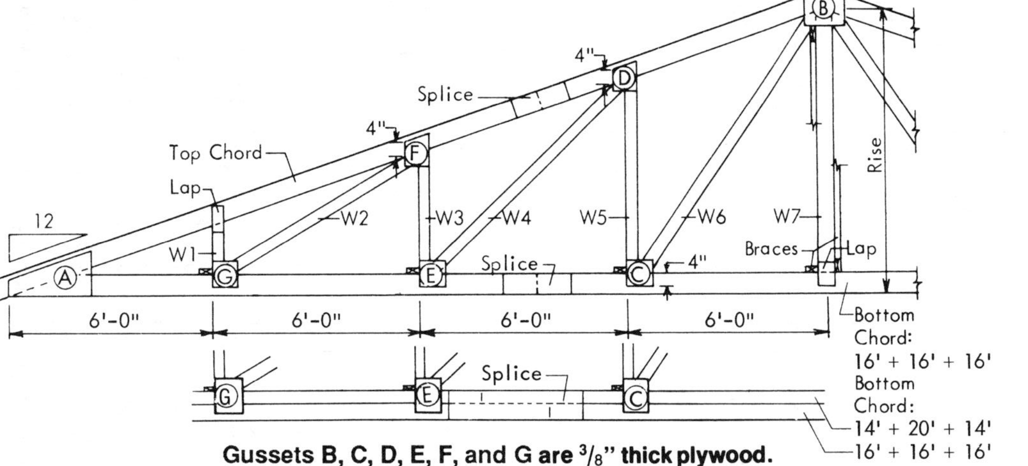

Gussets B, C, D, E, F, and G are 3/8" thick plywood.

4+4, 4+6, 6+6 indicates stacked lower chord.
4&4, 6&4, indicate double web; a 2x4 is attached to the web member to increase its stiffness.

Before selecting heel gusset A, see pages 8 and 9.

2 web lengths indicate double member, see page 10.

Web Lengths

Roof Slope	Rise	Top Chord	W1	W2	W3	W4	W5	W6	W7
3/12	6'-0"	16'+9"	2'	7'	3'	7'	5'	8'+7'	6'
4/12	8'-0"	16'+0'	2'	7'	4'	8'	6'+5'	10'+9'	8'
5/12	10'-0"	18'+9'	3'	8'	5'	10'+9'	8'+7'	12'+11'	10'

1100f Lumber

			Truss spacing, ft.									Web member sizes							Gusset Sizes, in.						
			2'			4'			8'										A	B	C	D	E	F	G
			Ceiling dead load, psf																						
	Top chord	Bottom chord	0	5	8	0	5	8	0	5	8	W1	W2	W3	W4	W5	W6	W7	T H W	H W	H W	H W	H W	H W	H W
			---Max. snow + roof dead load, psf---																						
3/12 Slope	2x4	2x4	21	17	14	0	0	0	0	0	0	2x4	2x4	2x4	2x4	2x4	2x4	2x4	3/8x3½x22	8x12	8x8	8x8	8x8	8x8	8x8
	2x6	2x4	22	19	16	0	0	0	0	0	0	"	"	"	"	"	"	"	3/8x4x22	10x12	"	"	"	"	"
	2x6	2x6	32	28	26	0	0	0	0	0	0	"	"	"	"	"	"	"	3/8x4x32	10x12	"	"	"	"	"
	2x8	2x6	34	30	27	0	0	0	0	0	0	2x4	2x4	2x4	2x4	2x4	2x4	2x4	3/8x4x33	12x12	"	"	"	"	"
	2x10	4+4	49	45	43	21	18	0	0	0	0	"	"	"	"	"	"	"	½x4x28	14x16	12x10	"	10x8	"	10x8
	2x12	4+6	63	58	56	27	23	21	0	0	0	"	"	"	"	"	"	"	½x4x36	16x16	14x10	"	12x10	"	12x8
	2x12	6+6	71	66	63	31	27	24	0	0	0	"	"	"	"	"	"	"	½x4x44	16x20	16x10	8x10	16x10	"	14x8
4/12 Slope	2x4	2x4	25	22	20	0	0	0	0	0	0	2x4	2x4	2x4	2x4	2x4	2x4	2x4	3/8x3½x21	8x12	8x8	8x8	8x8	8x8	8x8
	2x6	2x4	29	26	22	0	0	0	0	0	0	"	"	"	"	"	"	"	3/8x4x22	10x12	"	"	"	"	"
	2x6	2x6	43	40	37	19	15	0	0	0	0	"	"	"	"	"	"	"	½x4x19	"	10x8	"	"	"	"
	2x8	2x6	46	43	40	20	16	14	0	0	0	2x4	2x4	2x4	2x4	2x4	2x4	2x4	½x4x19	12x12	10x8	8x8	8x8	8x8	8x8
	2x10	4+4	63	60	58	27	24	17	0	0	0	"	"	"	"	2x6	"	"	½x4x25	14x16	12x12	8x10	10x10	"	10x8
	2x12	4+6	81	76	73	35	32	30	0	13	0	"	"	"	"	"	"	"	½x4x30	16x16	14x12	8x12	14x10	"	12x8
	2x12	6+6	93	87	83	40	37	34	20	16	12	"	"	"	"	2x8	4&4	"	½x4x36	18x16	18x16	8x14	16x10	"	14x8
5/12 Slope	2x4	2x4	28	25	23	0	0	0	0	0	0	2x4	2x4	2x4	2x4	2x4	2x4	2x4	3/8x3½x19	8x12	8x8	8x8	8x8	8x8	8x8
	2x6	2x4	37	32	27	0	0	0	0	0	0	"	"	"	"	2x6	"	"	3/8x4x21	10x12	8x10	8x10	"	"	"
	2x6	2x6	54	51	49	23	20	17	0	0	0	"	"	"	"	2x8	4&4	"	½x4x19	10x16	10x12	8x14	"	"	"
	2x8	2x6	57	53	51	25	21	19	0	0	0	2x4	2x4	2x4	2x4	2x8	4&4	2x4	½x4x19	12x16	10x12	8x12	8x8	8x8	8x8
	2x10	4+4	78	75	72	34	31	23	0	0	0	"	"	"	"	4&4	"	"	½x4x25	14x16	12x10	8x8	12x8	"	10x8
	2x12	4+6	100	94	91	43	41	38	21	18	0	"	"	2x6	4&4	"	"	"	½x4x30	16x20	16x10	8x10	14x12	8x10	12x8
	2x12	6+6	-	100+	100+	50	46	44	25	21	17	"	"	"	"	"	"	"	½x4x37	18x16	18x10	"	18x12	8x10	14x8

1400f Lumber

Slope	Top chord	Bottom chord	Truss spacing 2', Ceiling dead load 0 psf	2', 5	2', 8	4', 0	4', 5	4', 8	8', 0	8', 5	8', 8	W1	W2	W3	W4	W5	W6	W7	A THW	B HW	C HW	D HW	E HW	F HW	G HW
			---Max. snow + roof dead load, psf---									Web member sizes							Gusset Sizes, in.						
3/12 Slope	2x4	2x4	26	22	20	0	0	0	0	0	0	2x4	2x4	2x4	2x4	2x4	2x4	2x4	3/8x3½x28	8x12	8x8	8x8	8x8	8x8	8x8
	2x6	2x4	29	26	24	0	0	0	0	0	0	"	"	"	"	"	"	"	½x4x16	10x12	"	"	"	"	"
	2x6	2x6	42	38	35	18	14	0	0	0	0	"	"	"	"	"	"	"	½x4x23	10x16	8x10	"	"	"	"
	2x8	2x6	44	41	38	19	15	12	0	0	0	2x4	2x4	2x4	2x4	2x4	2x4	2x4	½x4x24	12x16	8x10	8x8	8x8	8x8	8x8
	2x10	4+4	64	59	57	27	24	21	0	0	0	"	"	"	"	"	"	"	½x4x31	14x16	12x10	8x10	10x10	"	10x8
	2x12	4+6	82	76	73	36	32	30	0	14	0	"	"	"	"	"	"	"	½x4x39	16x20	14x10	"	14x10	"	12x8
	2x12	6+6	89	81	77	38	34	31	19	14	0	"	"	"	"	"	4&4	"	½x4x44	18x16	16x12	10x10	16x10	"	14x8
4/12 Slope	2x4	2x4	30	27	25	0	0	0	0	0	0	2x4	2x4	2x4	2x4	2x4	2x4	2x4	½x3½x14	8x12	8x8	8x8	8x8	8x8	8x8
	2x6	2x4	39	35	32	0	12	0	0	0	0	"	"	"	"	"	"	"	½x4x16	10x12	"	"	"	"	"
	2x6	2x6	57	53	51	25	21	18	0	0	0	"	"	"	"	"	"	"	½x4x24	10x16	10x10	"	8x10	"	"
	2x8	2x6	60	56	54	26	22	20	0	0	0	2x4	2x4	2x4	2x4	2x4	2x4	2x4	½x4x24	12x16	10x10	8x8	8x10	8x8	8x8
	2x10	4+4	83	79	76	36	33	30	0	13	0	"	"	"	"	2x6	2x4	2x4	½x4x32	14x16	12x12	8x12	12x10	"	10x8
	2x12	4+6	100+	100	96	46	43	41	23	20	13	"	"	"	"	2x8	4&4	"	½x4x39	18x16	16x16	8x16	14x10	"	12x8
	2x12	6+6	–	–	100+	50	46	44	25	21	18	"	"	"	"	"	"	"	½x4x44	18x20	18x16	10x14	16x12	"	14x8
5/12 Slope	2x4	2x4	34	31	29	14	12	0	0	0	0	2x4	2x4	2x4	2x4	2x4	2x4	2x4	½x3½x13	8x12	8x8	8x8	8x8	8x8	8x8
	2x6	2x4	48	44	39	21	15	0	0	0	0	"	"	"	"	2x6	"	"	½x4x16	10x12	8x10	8x10	8x8	8x8	8x8
	2x6	2x6	66	62	59	28	25	23	0	0	0	"	"	"	"	2x8	"	"	½x4x22	10x16	10x12	8x12	10x8	"	"
	2x8	2x6	75	70	68	32	29	27	0	12	0	2x4	2x4	2x4	2x4	2x8	2x4	2x4	½x4x24	12x16	10x14	8x12	10x8	8x8	8x8
	2x10	4+4	100+	98	95	45	43	38	22	14	0	"	"	"	4&4	4&4	4&4	"	½x4x32	16x16	14x10	8x10	12x10	"	10x8
	2x12	4+6	–	–	100+	57	54	52	28	25	18	"	"	2x6	"	"	"	"	2-½x6x22	18x20	18x12	8x12	16x12	8x10	12x8
	2x12	6+6	–	–	–	61	56	53	30	26	23	"	"	"	"	"	"	"	½x4x40	"	20x12	"	18x12	"	14x8

1600f Lumber

Slope	Top chord	Bottom chord	Truss spacing 2', Ceiling dead load 0 psf	2', 5	2', 8	4', 0	4', 5	4', 8	8', 0	8', 5	8', 8	W1	W2	W3	W4	W5	W6	W7	A THW	B HW	C HW	D HW	E HW	F HW	G HW
			---Max. snow + roof dead load, psf---									Web member sizes							Gusset Sizes, in.						
3/12 Slope	2x4	2x4	31	28	26	0	0	0	0	0	0	2x4	2x4	2x4	2x4	2x4	2x4	2x4	½x3½x19	8x12	8x8	8x8	8x8	8x8	8x8
	2x6	2x4	35	32	30	0	0	0	0	0	0	"	"	"	"	"	"	"	½x4x19	10x12	8x10	"	"	"	"
	2x6	2x6	50	44	53	21	18	15	0	0	0	"	"	"	"	"	"	"	½x4x27	10x16	10x10	"	8x10	"	"
	2x8	2x6	49	47	47	23	19	16	0	0	0	2x4	2x4	2x4	2x4	2x4	2x4	2x4	½x4x28	12x16	10x10	8x8	8x10	8x8	8x8
	2x10	4+4	76	71	69	33	30	28	0	12	0	"	"	"	"	"	"	"	½x4x37	14x20	12x10	8x10	12x10	"	10x8
	2x12	4+6	99	91	88	43	40	37	21	18	14	"	"	"	"	"	"	"	½x4x46	18x20	16x12	10x10	14x10	"	12x8
	2x12	6+6	–	99	100+	47	43	40	23	19	16	"	"	"	"	"	"	"	½x4x53	"	18x12	10x12	16x12	"	14x8
4/12 Slope	2x4	2x4	37	34	32	16	13	0	0	0	0	2x4	2x4	2x4	2x4	2x4	2x4	2x4	½x3½x17	8x12	8x8	8x8	8x8	8x8	8x8
	2x6	2x4	46	44	40	20	15	0	0	0	0	"	"	"	"	"	"	"	½x4x19	10x12	10x8	"	"	"	"
	2x6	2x6	68	63	61	29	26	23	0	0	0	"	"	"	"	"	"	"	½x4x28	10x16	10x10	"	"	"	"
	2x8	2x6	72	67	64	31	28	26	0	12	0	2x4	2x4	2x4	2x4	2x4	2x4	2x4	½x4x28	12x16	10x10	8x8	8x8	8x8	8x8
	2x10	4+4	100	94	91	43	41	38	21	17	0	"	"	"	"	2x6	4&4	"	2-½x6x22	16x16	14x14	8x14	12x10	"	10x8
	2x12	4+6	–	100+	100+	55	52	50	27	24	20	"	"	"	"	"	"	"	2-½x6x28	18x20	18x14	10x12	16x12	"	12x8
	2x12	6+6	–	–	–	62	57	54	31	26	23	"	"	"	"	"	"	"	2-½x6x34	"	20x14	"	18x12	"	14x10
5/12 Slope	2x4	2x4	41	39	37	18	15	12	0	0	0	2x4	2x4	2x4	2x4	2x4	2x4	2x4	½x3½x15	8x12	8x8	8x8	8x8	8x8	8x8
	2x6	2x4	58	53	49	25	19	0	0	0	0	"	"	"	"	2x6	"	"	½x4x18	10x16	10x10	8x10	"	"	"
	2x6	2x6	81	76	73	35	32	30	17	14	0	"	"	"	"	2x8	4&4	"	2-½x6x14	10x20	10x14	8x12	10x10	"	"
	2x8	2x6	89	84	81	39	36	34	19	16	0	2x4	2x4	2x4	2x4	2x8	4&4	2x4	2-½x6x14	12x20	10x14	8x14	10x10	8x8	8x8
	2x10	4+4	100+	100+	100+	54	51	49	27	22	0	"	"	"	"	4&4	"	"	2-½x6x20	16x20	16x12	8x12	14x10	"	10x8
	2x12	4+6	–	–	–	68	64	62	34	31	27	"	"	2x6	"	"	"	"	2-½x6x22	10x20	18x12	10x10	16x12	8x10	12x8
	2x12	6+6	–	–	–	75	69	66	37	33	31	"	"	"	"	"	"	"	2-½x6x28	20x20	20x12	"	18x14	8x12	14x10

50' SPAN, 6-WEB

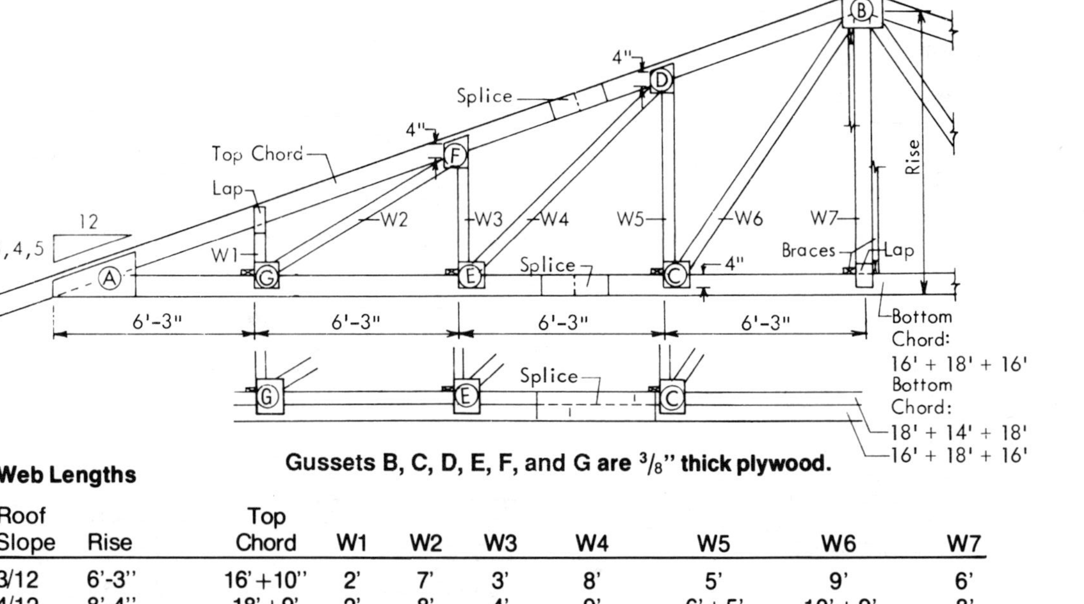

Gussets B, C, D, E, F, and G are $^{3}/_{8}$" thick plywood.

4+4, 4+6, 6+6 indicates stacked lower chord.
4&4, 6&4, indicate double web; a 2x4 is attached to the web member to increase its stiffness.

Before selecting heel gusset A, see pages 8 and 9 .

2 web lengths indicate double member, see page 10.

Web Lengths

Roof Slope	Rise	Top Chord	W1	W2	W3	W4	W5	W6	W7
3/12	6'-3"	16'+10"	2'	7'	3'	8'	5'	9'	6'
4/12	8'-4"	18'+9'	2'	8'	4'	9'	6'+5'	10'+9'	8'
5/12	10'-5"	18'+10'	3'	8'	5'	10'+9'	8'+7'	12'+11'	11'

1100f Lumber

	Top chord	Bottom chord	Truss spacing, ft. 2' — Ceiling dead load, psf 0	2' — 5	2' — 8	4' — 0	4' — 5	4' — 8	8' — 0	8' — 5	8' — 8	Web member sizes W1	W2	W3	W4	W5	W6	W7	Gusset Sizes, in. A T H W	B H W	C H W	D H W	E H W	F H W	G H W
			—Max. snow + roof dead load, psf—																						
3/12 Slope	2x4	2x4	0	16	12	0	0	0	0	0	0	2x4	2x4	2x4	2x4	2x4	2x4	2x4	3/8x3½x22	8x12	8x8	8x8	8x8	8x8	8x8
	2x6	2x4	0	18	15	0	0	0	0	0	0	"	"	"	"	"	"	"	3/8x4x22	10x12	"	"	"	"	"
	2x6	2x6	31	27	24	0	0	0	0	0	0	"	"	"	"	"	"	"	3/8x4x32	"	"	"	"	"	"
	2x8	2x6	32	28	26	0	0	0	0	0	0	2x4	2x4	2x4	2x4	2x4	2x4	2x4	3/8x4x34	12x12	"	"	"	"	"
	2x10	4+4	46	43	42	0	17	0	0	0	0	"	"	"	"	"	"	"	½x4x28	14x16	12x10	"	10x8	"	10x8
	2x12	4+6	60	56	53	26	22	20	0	0	0	"	"	"	"	"	"	"	½x4x36	16x16	14x10	8x10	12x10	"	12x8
	2x12	6+6	68	63	60	29	25	23	0	0	0	"	"	"	"	"	"	"	½x4x44	16x20	16x10	"	16x10	"	14x8
4/12 Slope	2x4	2x4	23	20	18	0	0	0	0	0	0	2x4	2x4	2x4	2x4	2x4	2x4	2x4	3/8x3½x21	8x12	8x8	8x8	8x8	8x8	8x8
	2x6	2x4	28	24	20	0	0	0	0	0	0	"	"	"	"	"	"	"	3/8x4x22	10x12	"	"	"	"	"
	2x6	2x6	42	39	36	18	14	0	0	0	0	"	"	"	"	"	4&4	"	½x4x19	"	10x8	"	"	"	"
	2x8	2x6	44	41	38	19	15	12	0	0	0	2x4	2x4	2x4	2x4	2x4	4&4	2x4	½x4x19	12x12	10x8	8x8	8x8	8x8	8x8
	2x10	4+4	61	57	55	26	23	15	0	0	0	"	"	"	"	2x6	"	"	½x4x25	14x16	12x12	8x10	10x10	"	10x8
	2x12	4+6	78	73	70	34	30	28	0	0	0	"	"	"	"	2x8	"	"	½x4x30	16x16	14x14	8x14	14x10	"	12x8
	2x12	6+6	90	83	80	39	35	33	19	15	12	"	"	"	4&4	"	"	"	½x4x36	18x16	18x16	8x16	16x10	"	14x8
5/12 Slope	2x4	2x4	26	23	22	0	0	0	0	0	0	2x4	2x4	2x4	2x4	2x4	2x4	2x4	3/8x3½x19	8x12	8x8	8x8	8x8	8x8	8x8
	2x6	2x4	35	30	25	0	0	0	0	0	0	"	"	"	"	2x6	"	"	3/8x4x21	10x12	8x10	8x10	"	"	"
	2x6	2x6	51	48	46	22	19	16	0	0	0	"	"	"	"	2x8	4&4	"	½x4x19	10x16	10x12	8x12	"	"	"
	2x8	2x6	55	51	50	24	21	16	0	0	0	2x4	2x4	2x4	2x4	2x8	4&4	2x4	½x4x19	12x16	10x12	8x12	8x8	8x8	8x8
	2x10	4+4	75	72	69	33	30	20	0	0	0	"	"	"	"	4&4	"	"	½x4x25	14x16	12x10	8x8	12x8	"	10x8
	2x12	4+6	96	90	87	41	39	36	20	16	0	"	"	2x6	4&4	"	"	"	½x4x31	16x20	16x10	8x10	14x12	8x10	12x8
	2x12	6+6	100+	100+	100+	48	44	41	24	20	14	"	"	"	"	"	"	"	½x4x37	18x16	18x10	8x12	18x12	"	14x8

1400f Lumber

Slope	Top chord	Bottom chord	2' spacing: 0	2': 5	2': 8	4' spacing: 0	4': 5	4': 8	8' spacing: 0	8': 5	8': 8	W1	W2	W3	W4	W5	W6	W7	A T H W	B H W	C H W	D H W	E H W	F H W	G H W
			Truss spacing, ft. / Ceiling dead load, psf / ---Max. snow + roof dead load, psf---									Web member sizes							Gusset Sizes, in.						
3/12 Slope	2x4	2x4	24	21	19	0	0	0	0	0	0	2x4	2x4	2x4	2x4	2x4	2x4	2x4	3/8x3½x28	8x12	8x8	8x8	8x8	8x8	8x8
	2x6	2x4	28	25	22	0	0	0	0	0	0	"	"	"	"	"	"	"	½x4x16	10x12	"	"	"	"	"
	2x6	2x6	40	36	34	0	13	0	0	0	0	"	"	"	"	"	"	"	½x4x23	10x16	8x10	"	"	"	"
	2x8	2x6	42	39	36	0	14	0	0	0	0	2x4	2x4	2x4	2x4	2x4	2x4	2x4	½x4x24	12x16	8x10	8x8	8x8	8x8	8x8
	2x10	4+4	61	57	55	26	23	19	0	0	0	"	"	"	"	"	"	"	½x4x32	14x16	12x10	8x10	10x10	"	10x8
	2x12	4+6	79	73	70	34	30	28	0	13	0	"	"	"	"	"	"	"	½x4x39	16x20	14x10	"	14x10	"	12x8
	2x12	6+6	85	78	74	37	32	30	0	14	0	"	"	"	"	"	"	"	½x4x44	18x16	16x12	10x10	16x10	"	14x8
4/12 Slope	2x4	2x4	28	25	24	0	0	0	0	0	0	2x4	2x4	2x4	2x4	2x4	2x4	2x4	½x3½x14	8x12	8x8	8x8	8x8	8x8	8x8
	2x6	2x4	37	34	29	0	0	0	0	0	0	"	"	"	"	"	"	"	½x4x16	10x12	10x12	"	"	"	"
	2x6	2x6	55	51	49	24	20	17	0	0	0	"	"	"	"	"	"	"	½x4x24	10x16	10x10	"	8x10	"	"
	2x8	2x6	58	54	52	25	21	19	0	0	0	2x4	2x4	2x4	2x4	2x6	2x4	2x4	½x4x24	12x16	10x12	8x10	8x10	8x8	8x8
	2x10	4+4	80	75	73	34	32	27	0	0	0	"	"	"	"	"	"	"	½x4x32	16x16	14x14	8x12	12x10	"	10x8
	2x12	4+6	100+	96	92	44	41	39	0	19	0	"	"	"	"	2x8	"	"	½x4x39	18x16	16x16	8x16	14x10	"	12x8
	2x12	6+6	–	100+	100+	48	44	41	24	20	17	"	"	"	4&4	"	4&4	"	½x4x45	18x20	18x16	10x14	16x12	"	14x8
5/12 Slope	2x4	2x4	32	29	27	0	0	0	0	0	0	2x4	2x4	2x4	2x4	2x4	2x4	2x4	3/8x3½x22	8x12	8x8	8x8	8x8	8x8	8x8
	2x6	2x4	46	41	37	0	14	0	0	0	0	"	"	"	"	2x6	"	"	½x4x16	10x12	8x10	8x10	"	"	"
	2x6	2x6	62	58	56	27	24	22	0	0	0	"	"	"	"	2x8	4&4	"	½x4x22	10x16	10x12	8x12	10x8	"	"
	2x8	2x6	72	67	65	31	28	26	0	0	0	2x4	2x4	2x4	2x4	4&4	4&4	2x4	½x4x24	12x16	10x10	8x8	10x8	8x8	8x8
	2x10	4+4	99	95	91	43	41	35	0	13	0	"	"	"	"	"	"	"	½x4x32	16x16	14x10	8x10	12x10	"	10x8
	2x12	4+6	–	100+	100+	54	51	50	27	24	15	"	"	2x6	4&4	"	"	"	2–½x6x22	18x20	18x12	8x12	16x12	8x10	12x8
	2x12	6+6	–	–	–	59	54	51	29	25	22	"	"	"	"	"	"	"	½x4x40	"	20x12	"	18x12	"	14x8

1600f Lumber

Slope	Top chord	Bottom chord	2' spacing: 0	2': 5	2': 8	4' spacing: 0	4': 5	4': 8	8' spacing: 0	8': 5	8': 8	W1	W2	W3	W4	W5	W6	W7	A T H W	B H W	C H W	D H W	E H W	F H W	G H W
			Truss spacing, ft. / Ceiling dead load, psf / ---Max. snow + roof dead load, psf---									Web member sizes							Gusset Sizes, in.						
3/12 Slope	2x4	2x4	30	27	25	0	0	0	0	0	0	2x4	2x4	2x4	2x4	2x4	2x4	2x4	½x3½x19	8x12	8x8	8x8	8x8	8x8	8x8
	2x6	2x4	33	30	28	0	0	0	0	0	0	"	"	"	"	"	"	"	½x4x19	10x12	8x10	"	"	"	"
	2x6	2x6	48	44	42	21	17	14	0	0	0	"	"	"	"	"	"	"	½x4x28	10x16	10x10	"	8x10	"	"
	2x8	2x6	51	47	45	22	18	15	0	0	0	2x4	2x4	2x4	2x4	2x4	2x4	2x4	½x4x29	12x16	10x10	8x8	8x10	8x8	8x8
	2x10	4+4	73	68	66	32	28	26	0	0	0	"	"	"	"	"	"	"	½x4x37	14x20	12x10	8x10	12x10	"	10x8
	2x12	4+6	94	87	84	41	37	35	0	17	0	"	"	"	"	"	"	"	½x4x46	18x20	16x12	10x10	14x10	"	12x8
	2x12	6+6	100+	95	97	45	41	38	22	18	15	"	"	"	"	"	4&4	"	½x4x54	"	18x12	10x12	16x12	"	14x8
4/12 Slope	2x4	2x4	35	32	30	0	12	0	0	0	0	2x4	2x4	2x4	2x4	2x4	2x4	2x4	½x3½x17	8x12	8x8	8x8	8x8	8x8	8x8
	2x6	2x4	44	42	38	0	14	0	0	0	0	"	"	"	"	"	"	"	½x4x19	10x12	10x8	"	"	"	"
	2x6	2x6	66	61	59	28	25	22	0	0	0	"	"	"	"	"	"	"	½x4x28	10x16	10x10	"	8x10	"	"
	2x8	2x6	69	64	62	30	26	24	0	0	0	2x4	2x4	2x4	2x4	2x4	2x4	2x4	½x4x28	12x16	10x10	8x8	8x10	8x8	8x8
	2x10	4+4	96	91	88	41	39	37	0	15	0	"	"	"	"	2x6	"	"	2–½x6x22	16x16	14x14	8x14	12x10	"	10x8
	2x12	4+6	100+	100+	100+	53	50	48	26	23	17	"	"	"	"	2x8	4&4	"	2–½x6x28	18x20	18x16	10x14	16x12	"	12x8
	2x12	6+6	–	–	–	59	54	52	29	25	22	"	"	"	"	2x6	"	"	2–½x6x34	"	20x14	"	18x12	"	14x10
5/12 Slope	2x4	2x4	39	37	34	17	14	0	0	0	0	2x4	2x4	2x4	2x4	2x4	2x4	2x4	½x3½x15	8x12	8x8	8x8	8x8	8x8	8x8
	2x6	2x4	55	51	47	24	18	0	0	0	0	"	"	"	"	"	4&4	"	½x4x19	10x16	10x8	"	"	"	"
	2x6	2x6	76	72	69	33	30	28	0	13	0	"	"	"	"	"	"	"	½x4x26	10x20	10x10	"	10x8	"	"
	2x8	2x6	86	81	78	37	34	32	0	15	0	2x4	2x4	2x4	2x4	4&4	2x4	2x4	2–½x6x14	12x20	10x10	8x10	10x10	8x8	8x8
	2x10	4+4	100+	100+	100+	51	49	47	25	19	0	"	"	"	4&4	"	4&4	"	2–½x6x20	16x20	16x12	8x12	14x10	"	10x8
	2x12	4+6	–	–	–	65	62	60	32	30	23	"	"	"	"	"	"	"	2–½x6x22	18x20	18x12	10x10	16x10	"	12x8
	2x12	6+6	–	–	–	72	66	63	36	32	29	"	"	2x6	"	"	"	"	2–½x6x28	20x20	20x12	"	18x14	8x12	14x10

52' SPAN, 6-WEB

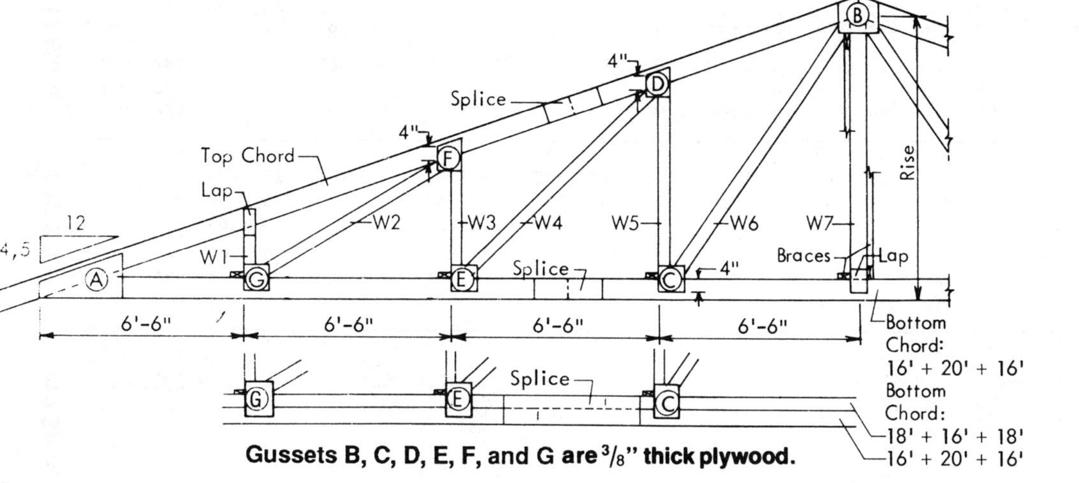

Gussets B, C, D, E, F, and G are 3/8" thick plywood.

4+4, 4+6, 6+6 indicates stacked lower chord.
4&4, 6&4, indicate double web; a 2x4 is attached to the web member to increase its stiffness.

Before selecting heel gusset A, see pages 8 and 9 .

2 web lengths indicate double member, see page 10.

Web Lengths

Roof Slope	Rise	Top Chord	W1	W2	W3	W4	W5	W6	W7
3/12	6'-6"	18'+10"	2'	7'	4'	8'	5'	9'+8'	7'
4/12	8'-8"	18'+10'	2'	8'	5'	10'+9'	7'+6'	11'+10'	9'
5/12	10'-10"	20'+9'	3'	9'	6'	11'+10'	8'+7'	13'+12'	11'

1100f Lumber

			Truss spacing, ft.									Web member sizes							Gusset Sizes, in.						
			2'			4'			8'										A	B	C	D	E	F	G
	Top chord	Bottom chord	Ceiling dead load, psf: 0	5	8	0	5	8	0	5	8	W1	W2	W3	W4	W5	W6	W7	T H W	H W	H W	H W	H W	H W	H W
			—Max. snow + roof dead load, psf—																						
3/12 Slope	2x4	2x4	0	15	0	0	0	0	0	0	0	2x4	2x4	2x4	2x4	2x4	2x4	2x4	3/8x3½x22	8x12	8x8	8x8	8x8	8x8	8x8
	2x6	2x4	0	17	12	0	0	0	0	0	0	"	"	"	"	"	"	"	3/8x4x22	10x12	"	"	"	"	"
	2x6	2x6	30	26	23	0	0	0	0	0	0	"	"	"	"	"	"	"	3/8x4x32	"	"	"	"	"	"
	2x8	2x6	31	27	24	0	0	0	0	0	0	2x4	2x4	2x4	2x4	2x4	2x4	2x4	3/8x4x33	12x12	8x8	8x8	8x8	8x8	8x8
	2x10	4+4	44	42	39	0	15	0	0	0	0	"	"	"	"	"	"	"	½x4x28	14x16	12x10	"	10x8	"	10x8
	2x12	4+6	57	53	51	25	21	18	0	0	0	"	"	"	"	"	"	"	½x4x36	16x16	14x10	8x10	12x10	"	12x8
	2x12	6+6	65	60	58	28	24	22	0	0	0	"	"	"	"	"	"	"	½x4x44	16x20	16x10	"	16x10	"	14x8
4/12 Slope	2x4	2x4	22	19	17	0	0	0	0	0	0	2x4	2x4	2x4	2x4	2x4	2x4	2x4	3/8x3½x21	8x12	8x8	8x8	8x8	8x8	8x8
	2x6	2x4	27	23	19	0	0	0	0	0	0	"	"	"	"	"	"	"	3/8x4x21	10x12	"	"	"	"	"
	2x6	2x6	40	37	34	0	13	0	0	0	0	"	"	"	"	2x6	"	"	½x4x19	"	10x10	8x10	"	"	"
	2x8	2x6	42	39	37	0	15	0	0	0	0	2x4	2x4	2x4	2x4	2x6	2x4	2x4	½x4x19	12x12	10x10	8x10	8x8	8x8	8x8
	2x10	4+4	58	55	53	25	22	12	0	0	0	"	"	"	"	"	4&4	"	½x4x25	14x16	12x12	8x10	10x10	"	10x8
	2x12	4+6	75	70	67	32	29	27	0	0	0	"	"	"	"	2x8	"	"	½x4x31	16x16	14x14	8x14	14x10	"	12x8
	2x12	6+6	86	80	77	37	34	31	0	14	0	"	"	"	4&4	"	"	"	½x4x36	18x16	18x16	8x16	16x10	"	14x8
5/12 Slope	2x4	2x4	24	22	20	0	0	0	0	0	0	2x4	2x4	2x4	2x4	2x4	4&4	2x4	3/8x3½x19	8x12	8x8	8x8	8x8	8x8	8x8
	2x6	2x4	34	28	24	0	0	0	0	0	0	"	"	"	"	2x6	"	"	3/8x4x21	10x12	8x10	8x10	"	"	"
	2x6	2x6	48	45	44	21	18	13	0	0	0	"	"	"	"	2x8	"	"	½x4x19	10x16	10x12	8x12	"	"	"
	2x8	2x6	53	50	48	23	20	13	0	0	0	2x4	2x4	2x4	2x4	2x8	4&4	2x4	½x4x19	12x16	10x12	8x12	8x8	8x8	8x8
	2x10	4+4	72	69	66	31	28	17	0	0	0	"	"	"	"	4&4	"	"	½x4x25	14x16	12x10	8x8	12x8	"	10x8
	2x12	4+6	92	87	84	40	38	35	0	13	0	"	"	2x6	"	"	"	"	½x4x31	16x20	16x10	8x10	14x12	8x10	12x8
	2x12	6+6	100+	98	100	46	43	40	23	19	12	"	"	"	4&4	"	"	"	½x4x37	18x20	18x10	8x12	18x12	8x10	14x8

1400f Lumber

Slope	Top chord	Bottom chord	Truss spacing, ft. 2' — Ceiling dead load, psf 0	2' 5	2' 8	4' 0	4' 5	4' 8	8' 0	8' 5	8' 8	Web member sizes W1	W2	W3	W4	W5	W6	W7	Gusset Sizes, in. A T H W	B H W	C H W	D H W	E H W	F H W	G H W
			---Max. snow + roof dead load, psf---																						
3/12 Slope	2x4	2x4	23	20	18	0	0	0	0	0	0	2x4	2x4	2x4	2x4	2x4	2x4	2x4	3/8x3½x28	8x12	8x8	8x8	8x8	8x8	8x8
	2x6	2x4	27	24	21	0	0	0	0	0	0	"	"	"	"	"	"	"	½x4x16	10x12	"	"	"	"	"
	2x6	2x6	39	35	33	0	12	0	0	0	0	"	"	"	"	"	"	"	½x4x24	10x16	8x10	"	"	"	"
	2x8	2x6	41	37	35	0	13	0	0	0	0	2x4	2x4	2x4	2x4	2x4	2x4	2x4	½x4x24	12x16	8x10	8x8	8x8	8x8	8x8
	2x10	4+4	58	55	53	25	22	17	0	0	0	"	"	"	"	"	"	"	½x4x32	14x16	12x10	8x10	10x10	"	10x8
	2x12	4+6	75	70	67	33	29	27	0	12	0	"	"	"	"	"	"	"	½x4x39	16x20	14x10	"	14x10	"	12x8
	2x12	6+6	82	75	71	35	31	28	0	13	0	"	"	"	"	"	"	"	½x4x44	18x16	16x12	10x10	16x10	"	14x8
4/12 Slope	2x4	2x4	27	24	22	0	0	0	0	0	0	2x4	2x4	2x4	2x4	2x4	2x4	2x4	½x3½x14	8x12	8x8	8x8	8x8	8x8	8x8
	2x6	2x4	36	32	28	0	0	0	0	0	0	"	"	"	"	"	"	"	½x4x16	10x12	"	"	"	"	"
	2x6	2x6	52	48	46	22	19	16	0	0	0	"	"	"	"	2x6	4&4	"	½x4x24	10x16	10x10	"	8x10	8x10	"
	2x8	2x6	55	52	50	24	21	18	0	0	0	2x4	2x4	2x4	2x4	2x6	4&4	2x4	½x4x24	12x16	10x10	8x8	8x10	8x10	8x8
	2x10	4+4	77	73	70	33	30	24	0	0	0	"	"	"	"	"	"	"	½x4x32	16x16	14x12	8x10	12x10	"	10x8
	2x12	4+6	98	92	88	42	40	37	0	18	0	"	"	"	"	2x8	"	"	½x4x40	18x16	16x12	8x12	14x10	8x12	12x8
	2x12	6+6	–	98	99	46	42	39	23	19	16	"	"	"	4&4	4&4	"	"	½x4x45	18x20	20x12	10x10	16x12	8x8	14x10
5/12 Slope	2x4	2x4	30	27	26	0	0	0	0	0	0	2x4	2x4	2x4	2x4	4&4	4&4	2x4	3/8x3½x22	8x12	8x8	8x8	8x8	8x8	8x8
	2x6	2x4	44	39	35	0	0	0	0	0	0	"	"	"	"	"	"	"	½x4x16	10x12	"	"	"	"	"
	2x6	2x6	59	55	53	25	22	20	0	0	0	"	"	"	"	"	"	"	½x4x22	10x16	10x8	"	10x8	"	"
	2x8	2x6	69	65	63	30	27	24	0	0	0	2x4	2x4	2x4	2x4	4&4	4&4	2x4	½x4x24	12x16	10x10	8x8	10x8	8x8	8x8
	2x10	4+4	95	91	92	41	39	31	0	0	0	"	"	2x6	"	"	"	"	2-½x6x18	16x16	14x10	8x10	14x12	8x10	10x8
	2x12	4+6	100+	100+	100+	52	50	48	26	23	12	"	"	"	4&4	"	"	"	2-½x6x22	18x20	18x12	8x12	16x12	"	12x8
	2x12	6+6	–	–	–	56	52	49	28	24	21	"	"	"	"	"	"	"	½x4x40	"	20x12	"	18x12	"	14x8

1600f Lumber

Slope	Top chord	Bottom chord	Truss spacing, ft. 2' — Ceiling dead load, psf 0	2' 5	2' 8	4' 0	4' 5	4' 8	8' 0	8' 5	8' 8	Web member sizes W1	W2	W3	W4	W5	W6	W7	Gusset Sizes, in. A T H W	B H W	C H W	D H W	E H W	F H W	G H W
			---Max. snow + roof dead load, psf---																						
3/12 Slope	2x4	2x4	28	25	23	0	0	0	0	0	0	2x4	2x4	2x4	2x4	2x4	2x4	2x4	½x3½x19	8x12	8x8	8x8	8x8	8x8	8x8
	2x6	2x4	32	29	27	0	0	0	0	0	0	"	"	"	"	"	"	"	½x4x20	10x12	8x10	"	"	"	"
	2x6	2x6	46	43	40	20	16	13	0	0	0	"	"	"	"	"	"	"	½x4x28	10x16	10x10	"	8x10	"	"
	2x8	2x6	49	45	43	21	17	14	0	0	0	2x4	2x4	2x4	2x4	2x4	2x4	2x4	½x4x29	12x16	10x10	8x8	8x10	8x8	8x8
	2x10	4+4	70	66	63	30	27	24	0	0	0	"	"	"	"	"	"	"	½x4x38	14x20	12x10	8x10	12x10	"	10x8
	2x12	4+6	90	84	80	39	35	33	0	16	0	"	"	"	"	"	"	"	½x4x46	18x20	16x12	10x10	14x10	"	12x8
	2x12	6+6	100+	92	93	44	39	36	22	17	14	"	"	"	"	"	4&4	"	½x4x54	"	18x12	10x12	16x12	"	14x8
4/12 Slope	2x4	2x4	33	30	28	0	0	0	0	0	0	2x4	2x4	2x4	2x4	2x4	2x4	2x4	½x3½x17	8x12	8x8	8x8	8x8	8x8	8x8
	2x6	2x4	43	40	36	0	13	0	0	0	0	"	"	"	"	"	"	"	½x4x19	10x12	10x8	"	"	"	"
	2x6	2x6	64	59	57	27	24	21	0	0	0	"	"	"	"	2x6	"	"	½x4x29	10x16	10x12	8x12	10x10	"	"
	2x8	2x6	66	62	60	29	25	23	0	0	0	2x4	2x4	2x4	2x4	2x6	2x4	2x4	½x4x28	12x16	10x12	8x12	10x10	8x8	8x8
	2x10	4+4	92	87	84	40	38	34	0	13	0	"	"	"	"	"	"	"	2-½x6x22	16x16	14x14	8x14	12x10	"	10x8
	2x12	4+6	100+	100+	100+	51	48	46	25	22	15	"	"	"	"	2x8	4&4	"	2-½x6x38	18x20	18x16	10x14	16x12	"	12x8
	2x12	6+6	–	–	–	57	52	50	28	24	21	"	"	"	"	"	"	"	2-½x6x34	"	20x16	10x16	18x12	"	14x10
5/12 Slope	2x4	2x4	37	34	32	16	13	0	0	0	0	2x4	2x4	2x4	2x4	4&4	4&4	2x4	½x3½x15	8x12	8x8	8x8	8x8	8x8	8x8
	2x6	2x4	53	49	44	23	17	0	0	0	0	"	"	"	"	"	"	"	½x4x19	10x16	10x8	"	"	"	"
	2x6	2x6	72	68	65	31	28	26	0	12	0	"	"	"	"	"	"	"	½x4x26	10x20	10x10	"	10x8	"	"
	2x8	2x6	83	78	75	36	33	31	0	14	0	2x4	2x4	2x4	2x4	4&4	4&4	2x4	2-½x6x14	12x20	12x10	8x10	10x10	8x8	8x8
	2x10	4+4	100+	100+	100+	49	48	43	24	16	0	"	"	"	4&4	"	"	"	2-½x6x20	16x20	16x12	8x12	14x10	"	10x8
	2x12	4+6	–	–	–	63	60	58	31	28	20	"	"	2x6	"	"	"	"	2-½x6x23	18x20	18x12	10x10	16x14	8x10	12x8
	2x12	6+6	–	–	–	69	64	61	35	30	28	"	"	"	"	"	"	"	2-½x6x28	20x20	20x12	10x12	18x14	8x12	14x10

54' SPAN, 6-WEB

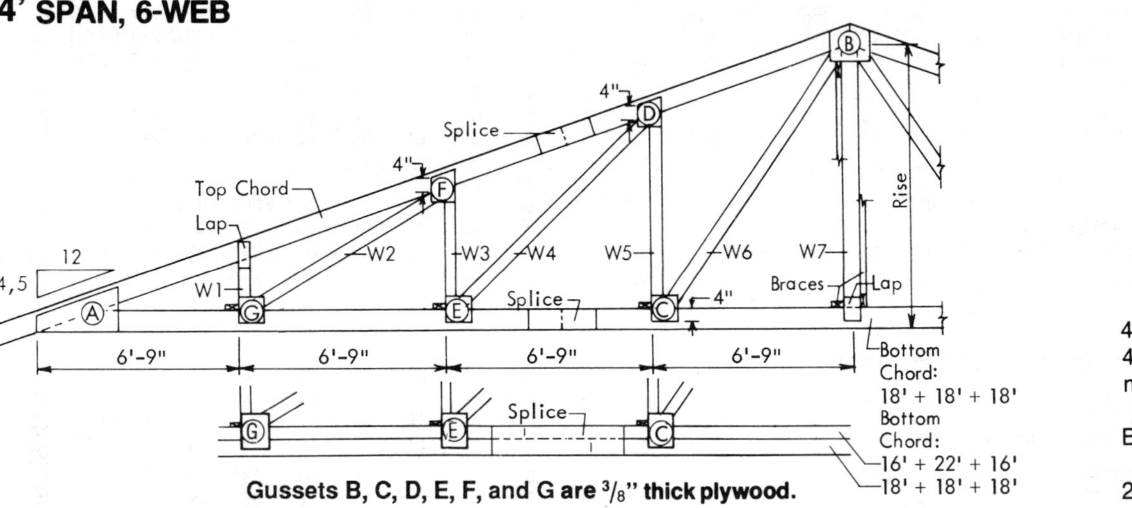

Gussets B, C, D, E, F, and G are $^3/_8$" thick plywood.

4+4, 4+6, 6+6 indicates stacked lower chord.
4&4, 6&4, indicate double web; a 2x4 is attached to the web member to increase its stiffness.

Before selecting heel gusset A, see pages 8 and 9 .

2 web lengths indicate double member, see page 10.

Web Lengths

Roof Slope	Rise	Top Chord	W1	W2	W3	W4	W5	W6	W7
3/12	6'-9"	18'+11"	2'	7'	4'	8'	5'	10'+9'	7'
4/12	9'-0"	18'+11'	2'	8'	5'	10'	7'+6'	11'+10'	9'
5/12	11'-3"	20'+10'	3'	9'	6'	11'+10'	9'+8'	13'+12'	11'+10'

1100f Lumber

			Truss spacing, ft.																						
			2'			4'			8'										Gusset Sizes, in.						
			Ceiling dead load, psf									Web member sizes							A	B	C	D	E	F	G
	Top chord	Bottom chord	0	5	8	0	5	8	0	5	8	W1	W2	W3	W4	W5	W6	W7	T H W	H W	H W	H W	H W	H W	H W
			---Max. snow + roof dead load, psf---																						
3/12 Slope	2x4	2x4	0	14	0	0	0	0	0	0	0	2x4	2x4	2x4	2x4	2x4	2x4	2x4	3/8x3½x22	8x12	8x8	8x8	8x8	8x8	8x8
	2x6	2x4	0	16	0	0	0	0	0	0	0	"	"	"	"	"	"	"	3/8x4x22	10x12	"	"	"	"	"
	2x6	2x6	29	25	22	0	0	0	0	0	0	"	"	"	"	"	"	"	3/8x4x33	"	"	"	"	"	"
	2x8	2x6	30	26	23	0	0	0	0	0	0	2x4	2x4	2x4	2x4	2x4	2x4	2x4	3/8x4x34	12x12	8x8	8x8	8x8	8x8	8x8
	2x10	4+4	43	40	37	0	15	0	0	0	0	"	"	"	"	"	"	"	½x4x28	14x16	12x10	"	10x8	"	10x8
	2x12	4+6	55	51	49	0	20	17	0	0	0	"	"	"	"	"	"	"	½x4x36	16x16	14x10	8x10	12x10	"	12x8
	2x12	6+6	63	58	56	27	23	21	0	0	0	"	"	"	"	"	"	"	½x4x44	16x20	16x10	"	16x10	"	14x8
4/12 Slope	2x4	2x4	21	18	15	0	0	0	0	0	0	2x4	2x4	2x4	2x4	2x4	2x4	2x4	3/8x3½x21	8x12	8x8	8x8	8x8	8x8	8x8
	2x6	2x4	26	21	15	0	0	0	0	0	0	"	"	"	"	"	"	"	3/8x4x21	10x12	"	"	"	"	"
	2x6	2x6	39	35	33	0	12	0	0	0	0	"	"	"	"	2x6	"	"	½x4x19	"	10x10	8x10	"	"	"
	2x8	2x6	41	38	35	0	14	0	0	0	0	2x4	2x4	2x4	2x4	2x6	2x4	2x4	½x4x19	12x12	10x10	8x10	8x8	8x8	8x8
	2x10	4+4	56	53	51	24	20	0	0	0	0	"	"	"	"	"	4&4	"	½x4x26	14x16	12x12	"	10x10	"	10x8
	2x12	4+6	72	67	65	31	28	25	0	0	0	"	"	"	"	2x8	"	"	½x4x31	16x16	14x14	8x14	14x10	"	12x8
	2x12	6+6	83	77	74	36	32	30	0	14	0	"	"	"	"	4&4	"	"	½x4x37	18x16	18x12	8x12	16x10	"	14x8
5/12 Slope	2x4	2x4	23	21	19	0	0	0	0	0	0	2x4	2x4	2x4	2x4	4&4	4&4	2x4	3/8x3½x19	8x12	8x8	8x8	8x8	8x8	8x8
	2x6	2x4	33	27	20	0	0	0	0	0	0	"	"	"	"	"	"	"	3/8x4x22	10x12	"	"	"	"	"
	2x6	2x6	46	43	41	20	17	0	0	0	0	"	"	"	4&4	"	"	"	½x4x19	10x16	10x8	"	"	"	"
	2x8	2x6	51	48	46	22	19	14	0	0	0	2x4	2x4	2x4	4&4	4&4	4&4	2x4	½x4x19	12x16	8x8	8x8	8x8	8x8	8x8
	2x10	4+4	70	67	63	30	26	13	0	0	0	"	"	2x6	"	"	"	"	½x4x25	14x20	12x10	"	12x10	8x10	10x8
	2x12	4+6	89	84	81	38	36	32	0	12	0	"	"	"	"	"	"	"	½x4x31	16x20	16x10	8x10	14x12	"	12x8
	2x12	6+6	100+	95	96	45	41	38	0	18	0	"	"	"	"	6&4	"	"	½x4x37	18x20	18x12	8x14	18x12	"	14x8

1400f Lumber

Slope	Top chord	Bottom chord	2' 0	2' 5	2' 8	4' 0	4' 5	4' 8	8' 0	8' 5	8' 8	W1	W2	W3	W4	W5	W6	W7	A T H W	B H W	C H W	D H W	E H W	F H W	G H W
			Truss spacing, ft. / Ceiling dead load, psf / ---Max. snow + roof dead load, psf---									Web member sizes							Gusset Sizes, in.						
3/12 Slope	2x4	2x4	0	19	16	0	0	0	0	0	0	2x4	2x4	2x4	2x4	2x4	2x4	2x4	3/8x3½x28	8x12	8x8	8x8	8x8	8x8	8x8
	2x6	2x4	0	23	19	0	0	0	0	0	0	"	"	"	"	"	"	"	½x4x16	10x12	"	"	"	"	"
	2x6	2x6	37	34	31	0	12	0	0	0	0	"	"	"	"	"	"	"	½x4x24	10x16	8x10	"	"	"	"
	2x8	2x6	39	35	33	0	13	0	0	0	0	2x4	2x4	2x4	2x4	2x4	2x4	2x4	½x4x25	12x16	10x10	8x8	8x8	8x8	8x8
	2x10	4+4	56	52	51	0	21	15	0	0	0	"	"	"	"	"	"	"	½x4x33	14x16	12x10	8x10	10x10	"	10x8
	2x12	4+6	72	67	64	31	28	26	0	0	0	"	"	"	"	"	"	"	½x4x39	16x20	14x10	"	14x10	"	12x8
	2x12	6+6	79	72	68	34	30	27	0	12	0	"	"	"	"	"	"	"	½x4x45	18x16	16x12	10x10	16x10	"	14x8
4/12 Slope	2x4	2x4	25	23	21	0	0	0	0	0	0	2x4	2x4	2x4	2x4	2x4	2x4	2x4	½x3½x14	8x12	8x8	8x8	8x8	8x8	8x8
	2x6	2x4	34	30	26	0	0	0	0	0	0	"	"	"	"	"	"	"	½x4x16	10x12	"	"	"	"	"
	2x6	2x6	49	46	44	0	18	15	0	0	0	"	"	"	"	2x6	"	"	½x4x24	10x16	10x12	8x10	8x10	"	"
	2x8	2x6	53	50	48	0	20	17	0	0	0	2x4	2x4	2x4	2x4	2x6	2x4	2x4	½x4x25	12x16	10x12	8x10	8x10	8x8	8x8
	2x10	4+4	74	70	67	32	29	21	0	0	0	"	"	"	"	2x8	4&4	"	½x4x33	16x16	14x16	8x14	12x10	"	10x8
	2x12	4+6	94	88	85	41	38	36	0	16	0	"	"	"	"	4&4	"	"	½x4x40	18x16	16x12	8x12	14x10	"	12x8
	2x12	6+6	100+	95	95	45	41	38	0	18	15	"	"	"	"	"	"	"	½x4x45	18x20	20x12	10x10	16x12	"	14x10
5/12 Slope	2x4	2x4	28	26	24	0	0	0	0	0	0	2x4	2x4	2x4	2x4	4&4	4&4	2x4	3/8x3½x22	8x12	8x8	8x8	8x8	8x8	8x8
	2x6	2x4	43	37	33	0	0	0	0	0	0	"	"	"	"	"	"	"	½x4x16	10x12	"	"	"	"	"
	2x6	2x6	56	52	50	24	21	19	0	0	0	"	"	"	"	"	"	"	½x4x22	10x16	10x8	"	10x8	"	"
	2x8	2x6	67	63	61	29	26	23	0	0	0	2x4	2x4	2x4	2x4	4&4	4&4	2x4	½x4x25	12x16	10x10	8x8	10x8	8x8	8x8
	2x10	4+4	92	88	88	40	37	27	0	0	0	"	"	2x6	"	"	"	"	2-½x6x18	16x16	14x10	8x10	14x12	8x10	10x8
	2x12	4+6	100+	100+	100+	50	48	46	0	21	0	"	"	"	"	6&4	"	"	2-½x6x22	18x20	18x14	8x14	16x12	"	12x8
	2x12	6+6	-	-	-	54	50	47	27	23	20	"	"	"	4&4	4&4	"	"	2-½x6x28	"	20x12	"	18x12	"	14x8

1600f Lumber

Slope	Top chord	Bottom chord	2' 0	2' 5	2' 8	4' 0	4' 5	4' 8	8' 0	8' 5	8' 8	W1	W2	W3	W4	W5	W6	W7	A T H W	B H W	C H W	D H W	E H W	F H W	G H W
			Truss spacing, ft. / Ceiling dead load, psf / ---Max. snow + roof dead load, psf---									Web member sizes							Gusset Sizes, in.						
3/12 Slope	2x4	2x4	27	24	22	0	0	0	0	0	0	2x4	2x4	2x4	2x4	2x4	2x4	2x4	½x3½x19	8x12	8x8	8x8	8x8	8x8	8x8
	2x6	2x4	31	28	25	0	0	0	0	0	0	"	"	"	"	"	"	"	½x4x19	10x12	8x10	"	"	"	"
	2x6	2x6	45	42	39	0	15	12	0	0	0	"	"	"	"	"	"	"	½x4x30	10x16	10x10	"	8x10	"	"
	2x8	2x6	47	44	41	0	16	14	0	0	0	2x4	2x4	2x4	2x4	2x4	2x4	2x4	½x4x29	12x16	10x10	8x8	8x10	8x8	8x8
	2x10	4+4	67	63	61	29	26	23	0	0	0	"	"	"	"	"	"	"	½x4x38	14x20	12x10	8x10	12x10	"	10x8
	2x12	4+6	86	80	77	37	34	32	0	15	0	"	"	"	"	"	"	"	½x4x47	18x20	16x12	10x10	14x12	"	12x8
	2x12	6+6	97	88	89	42	37	35	0	16	14	"	"	"	"	"	"	"	½x4x55	18x20	18x12	10x12	16x12	"	14x8
4/12 Slope	2x4	2x4	31	28	27	0	0	0	0	0	0	2x4	2x4	2x4	2x4	2x4	2x4	2x4	½x3½x17	8x12	8x8	8x8	8x8	8x8	8x8
	2x6	2x4	41	38	34	0	0	0	0	0	0	"	"	"	"	"	"	"	½x4x19	10x12	10x8	"	"	"	"
	2x6	2x6	61	57	55	26	23	20	0	0	0	"	"	"	"	2x6	"	"	½x4x29	10x16	10x10	8x12	10x10	"	"
	2x8	2x6	64	60	57	27	24	22	0	0	0	2x4	2x4	2x4	2x4	2x6	2x4	2x4	½x4x29	12x16	10x12	8x12	10x10	8x8	8x8
	2x10	4+4	88	84	81	38	36	31	0	0	0	"	"	"	"	"	"	"	2-½x6x22	16x16	14x14	8x14	12x10	"	10x8
	2x12	4+6	100+	100+	100+	49	46	44	24	21	13	"	"	"	4&4	2x8	4&4	"	2-½x6x28	18x20	18x16	10x16	16x12	"	12x8
	2x12	6+6	-	-	-	55	50	48	27	23	20	"	"	"	"	"	"	"	2-½x6x34	18x20	20x16	"	18x12	"	14x10
5/12 Slope	2x4	2x4	35	32	30	0	12	0	0	0	0	2x4	2x4	2x4	2x4	4&4	4&4	2x4	½x3½x15	8x12	8x8	8x8	8x8	8x8	8x8
	2x6	2x4	51	47	42	0	16	0	0	0	0	"	"	"	"	"	"	"	½x4x19	10x16	10x8	"	"	"	"
	2x6	2x6	68	64	62	29	26	25	0	0	0	"	"	"	"	"	"	"	½x4x26	10x20	10x10	"	10x8	"	"
	2x8	2x6	80	75	73	34	32	30	0	12	0	2x4	2x4	2x4	2x4	4&4	4&4	2x4	2-½x6x15	12x20	12x10	8x10	10x10	8x8	8x8
	2x10	4+4	100+	100+	100+	48	46	39	0	15	0	"	"	2x6	"	"	"	"	2-½x6x20	16x20	16x12	8x12	14x12	8x10	10x8
	2x12	4+6	-	-	-	60	57	55	30	27	17	"	"	"	4&4	"	"	"	2-½x6x23	18x20	18x12	10x10	16x14	"	12x8
	2x12	6+6	-	-	-	67	62	59	33	29	27	"	"	"	"	"	"	"	2-½x6x28	20x20	20x12	10x12	18x14	8x12	14x10

56' SPAN, 6-WEB

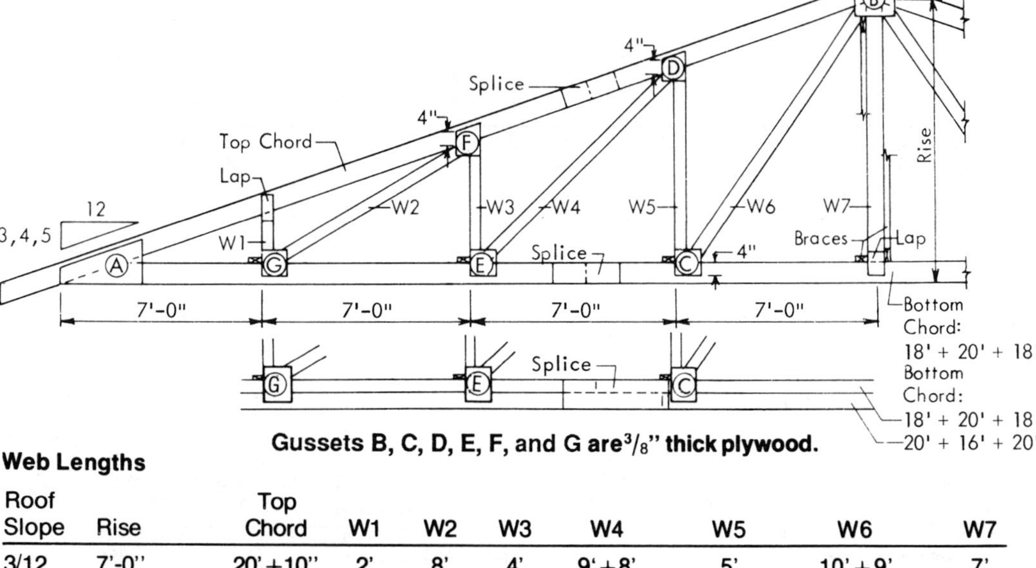

Gussets B, C, D, E, F, and G are 3/8" thick plywood.

4+4, 4+6, 6+6 indicates stacked lower chord.
4&4, 6&4, indicate double web; a 2x4 is attached to the web member to increase its stiffness.

Before selecting heel gusset A, see pages 8 and 9.

2 web lengths indicate double member, see page 10.

Web Lengths

Roof Slope	Rise	Top Chord	W1	W2	W3	W4	W5	W6	W7
3/12	7'-0"	20'+10"	2'	8'	4'	9'+8'	5'	10'+9'	7'
4/12	9'-4"	20'+10'	2'	8'	5'	10'+9'	7'+6'	12'+11'	9'
5/12	11'-8"	20'+11'	3'	9'	6'	11'+10'	9'+8'	13'+12'	12'+11'

1100f Lumber

			Truss spacing, ft.									Web member sizes							Gusset Sizes, in.						
			2'			4'			8'										A	B	C	D	E	F	G
			Ceiling dead load, psf																						
	Top chord	Bottom chord	0	5	8	0	5	8	0	5	8	W1	W2	W3	W4	W5	W6	W7	T H W	H W	H W	H W	H W	H W	H W
			—Max. snow + roof dead load, psf—																						
3/12 Slope	2x4	2x4	0	14	0	0	0	0	0	0	0	2x4	2x4	2x4	2x4	2x4	2x4	2x4	3/8x3½x22	8x12	8x8	8x8	8x8	8x8	8x8
	2x6	2x4	0	15	0	0	0	0	0	0	0	"	"	"	"	"	"	"	3/8x4x22	10x12	"	"	"	"	"
	2x6	2x6	28	24	21	0	0	0	0	0	0	"	"	"	"	"	"	"	3/8x4x33	"	"	"	"	"	"
	2x8	2x6	29	25	22	0	0	0	0	0	0	2x4	2x4	2x4	2x4	2x4	2x4	2x4	3/8x4x34	12x12	8x8	8x8	8x8	8x8	8x8
	2x10	4+4	41	38	36	0	13	0	0	0	0	"	"	"	"	"	"	"	½x4x28	14x16	12x10	"	10x8	"	10x8
	2x12	4+6	53	49	47	0	19	15	0	0	0	"	"	"	"	"	"	"	½x4x36	16x16	14x10	8x10	12x10	"	12x8
	2x12	6+6	60	56	53	26	22	20	0	0	0	"	"	"	"	"	4&4	"	½x4x44	16x20	16x10	"	16x10	"	14x8
4/12 Slope	2x4	2x4	0	17	0	0	0	0	0	0	0	2x4	2x4	2x4	2x4	2x4	2x4	2x4	3/8x3½x20	8x12	8x8	8x8	8x8	8x8	8x8
	2x6	2x4	0	20	0	0	0	0	0	0	0	"	"	"	"	"	"	"	3/8x4x21	10x12	"	"	"	"	"
	2x6	2x6	38	34	31	0	12	0	0	0	0	"	"	"	"	2x6	"	"	½x4x19	"	10x10	8x10	"	"	"
	2x8	2x6	39	36	34	0	13	0	0	0	0	2x4	2x4	2x4	2x4	2x6	2x4	2x4	½x4x19	12x12	10x10	8x10	8x8	8x8	8x8
	2x10	4+4	54	51	49	0	19	0	0	0	0	"	"	"	"	2x8	"	"	½x4x26	14x16	12x14	8x12	10x10	"	10x8
	2x12	4+6	69	65	63	30	27	23	0	0	0	"	"	"	4&4	"	4&4	"	½x4x31	16x16	14x14	8x14	14x10	"	12x8
	2x12	6+6	80	75	71	35	31	29	0	13	0	"	"	"	"	4&4	"	"	½x4x37	18x16	18x12	8x12	16x10	"	14x8
5/12 Slope	2x4	2x4	22	19	17	0	0	0	0	0	0	2x4	2x4	2x4	2x4	4&4	4&4	2x4	3/8x3½x18	8x12	8x8	8x8	8x8	8x8	8x8
	2x6	2x4	31	25	17	0	0	0	0	0	0	"	"	"	"	"	"	"	3/8x4x22	10x12	"	"	"	"	"
	2x6	2x6	43	41	38	0	16	0	0	0	0	"	"	"	"	"	"	"	½x4x18	10x16	10x8	"	"	"	"
	2x8	2x6	49	46	45	0	18	0	0	0	0	2x4	2x4	2x4	2x4	4&4	4&4	2x4	½x4x20	12x16	10x8	8x8	8x8	8x8	8x8
	2x10	4+4	67	64	60	29	24	0	0	0	0	"	"	2x6	4&4	"	"	"	½x4x26	14x20	12x10	"	12x10	8x10	10x8
	2x12	4+6	85	81	78	37	34	29	0	0	0	"	"	"	"	"	"	"	½x4x31	18x16	16x10	8x10	14x12	"	12x8
	2x12	6+6	100	92	92	43	40	36	0	17	0	"	"	2x8	"	"	"	"	½x4x37	18x20	18x10	8x12	18x14	8x12	14x8

1400f Lumber

Slope	Top chord	Bottom chord	Truss spacing 2', Ceiling dead load 0 psf	2', 5	2', 8	4', 0	4', 5	4', 8	8', 0	8', 5	8', 8	W1	W2	W3	W4	W5	W6	W7	A T H W	B H W	C H W	D H W	E H W	F H W	G H W
			--Max. snow + roof dead load, psf---									Web member sizes							Gusset Sizes, in.						
3/12 Slope	2x4	2x4	0	17	15	0	0	0	0	0	0	2x4	2x4	2x4	2x4	2x4	2x4	2x4	3/8x3½x28	8x12	8x8	8x8	8x8	8x8	8x8
	2x6	2x4	0	22	18	0	0	0	0	0	0	"	"	"	"	"	"	"	½x4x16	10x12	"	"	"	"	"
	2x6	2x6	36	32	30	0	0	0	0	0	0	"	"	"	"	"	"	"	½x4x24	10x16	8x10	"	"	"	"
	2x8	2x6	38	34	32	0	12	0	0	0	0	2x4	2x4	2x4	2x4	2x4	2x4	2x4	½x4x25	12x16	10x10	8x8	8x8	8x8	8x8
	2x10	4+4	54	51	49	0	20	12	0	0	0	"	"	"	"	"	"	"	½x4x33	14x20	12x10	8x10	10x10	"	10x8
	2x12	4+6	69	65	62	30	26	24	0	0	0	"	"	"	"	"	"	"	½x4x40	16x20	14x10	"	14x10	"	12x8
	2x12	6+6	76	69	66	33	28	25	0	12	0	"	"	"	"	2x6	"	"	½x4x45	18x16	16x14	10x12	16x10	"	14x8
4/12 Slope	2x4	2x4	24	21	20	0	0	0	0	0	0	2x4	2x4	2x4	2x4	2x4	2x4	2x4	½x3½x14	8x12	8x8	8x8	8x8	8x8	8x8
	2x6	2x4	33	29	25	0	0	0	0	0	0	"	"	"	"	"	"	"	½x4x16	10x12	"	"	"	"	"
	2x6	2x6	47	44	42	0	17	14	0	0	0	"	"	"	"	2x6	"	"	½x4x24	10x16	10x12	8x10	8x10	"	"
	2x8	2x6	51	48	46	0	19	15	0	0	0	2x4	2x4	2x4	2x4	2x6	2x4	2x4	½x4x25	12x16	10x12	8x10	8x10	8x8	8x8
	2x10	4+4	71	68	65	31	28	19	0	0	0	"	"	"	"	2x8	4&4	"	½x4x33	16x16	14x16	8x14	12x10	"	10x8
	2x12	4+6	91	85	82	39	37	34	0	15	0	"	"	"	"	4&4	"	"	½x4x40	18x16	16x12	8x12	14x10	"	12x8
	2x12	6+6	100	91	92	43	39	36	0	17	14	"	"	"	"	"	"	"	½x4x46	18x20	20x12	10x10	16x12	"	14x10
5/12 Slope	2x4	2x4	27	24	23	0	0	0	0	0	0	2x4	2x4	2x4	2x4	4&4	4&4	2x4	3/8x3½x22	8x12	8x8	8x8	8x8	8x8	8x8
	2x6	2x4	41	36	31	0	0	0	0	0	0	"	"	"	"	"	"	"	½x4x16	10x12	"	"	"	"	"
	2x6	2x6	53	50	48	23	20	18	0	0	0	"	"	"	4&4	"	"	"	½x4x22	10x16	10x8	"	10x8	"	"
	2x8	2x6	64	61	59	28	25	21	0	0	0	2x4	2x4	2x4	4&4	4&4	4&4	2x4	½x4x25	12x20	10x10	8x8	10x8	8x8	8x8
	2x10	4+4	88	85	85	38	36	24	0	0	0	"	"	2x6	"	"	"	"	2-½x6x18	16x16	14x10	8x10	14x12	8x10	10x8
	2x12	4+6	100+	100+	100+	49	46	44	0	20	0	"	"	2x8	"	"	"	"	2-½x6x22	18x20	18x12	8x12	16x14	8x12	12x8
	2x12	6+6	-	-	-	52	48	46	26	22	19	"	4&4	2x6	"	6&4	"	"	2-½x6x28	"	20x14	8x14	18x12	8x10	14x8

1600f Lumber

Slope	Top chord	Bottom chord	Truss spacing 2', Ceiling dead load 0 psf	2', 5	2', 8	4', 0	4', 5	4', 8	8', 0	8', 5	8', 8	W1	W2	W3	W4	W5	W6	W7	A T H W	B H W	C H W	D H W	E H W	F H W	G H W
			--Max. snow + roof dead load, psf---									Web member sizes							Gusset Sizes, in.						
3/12 Slope	2x4	2x4	25	22	21	0	0	0	0	0	0	2x4	2x4	2x4	2x4	2x4	2x4	2x4	½x3½x19	8x12	8x8	8x8	8x8	8x8	8x8
	2x6	2x4	30	27	24	0	0	0	0	0	0	"	"	"	"	"	"	"	½x4x19	10x12	8x10	"	"	"	"
	2x6	2x6	43	40	37	0	14	0	0	0	0	"	"	"	"	"	"	"	½x4x28	10x16	10x10	"	8x10	"	"
	2x8	2x6	45	42	39	0	15	13	0	0	0	2x4	2x4	2x4	2x4	2x4	2x4	2x4	½x4x29	12x16	10x10	8x8	8x10	8x8	8x8
	2x10	4+4	65	61	59	28	25	20	0	0	0	"	"	"	"	"	"	"	½x4x39	14x20	12x10	8x10	12x10	"	10x8
	2x12	4+6	83	77	74	36	32	30	0	14	0	"	"	"	"	"	"	"	½x4x47	18x20	16x12	10x10	14x12	"	12x8
	2x12	6+6	93	85	86	40	36	33	0	15	13	"	"	"	"	"	"	"	½x4x55	"	18x12	10x12	16x12	"	14x8
4/12 Slope	2x4	2x4	29	27	25	0	0	0	0	0	0	2x4	2x4	2x4	2x4	2x4	2x4	2x4	½x3½x16	8x12	8x8	8x8	8x8	8x8	8x8
	2x6	2x4	40	36	32	0	0	0	0	0	0	"	"	"	"	"	"	"	½x4x19	10x12	10x8	"	"	"	"
	2x6	2x6	57	54	52	25	22	19	0	0	0	"	"	"	"	2x6	"	"	½x4x29	10x16	10x12	8x12	10x10	"	"
	2x8	2x6	62	58	56	27	23	21	0	0	0	2x4	2x4	2x4	2x4	2x6	2x4	2x4	½x4x29	12x16	10x12	8x12	10x10	8x8	8x8
	2x10	4+4	85	81	78	37	34	29	0	0	0	"	"	"	"	2x8	"	"	2-½x6x22	16x16	14x16	8x16	12x10	"	10x8
	2x12	4+6	100+	100+	99	47	44	43	0	20	0	"	"	"	"	4&4	"	"	2-½x6x28	18x20	18x12	10x12	16x12	"	12x8
	2x12	6+6	-	-	-	53	49	46	26	22	19	"	"	"	4&4	"	4&4	"	2-½x6x34	20x20	20x12	"	18x12	8x10	14x10
5/12 Slope	2x4	2x4	33	30	29	0	0	0	0	0	0	2x4	2x4	2x4	2x4	4&4	4&4	2x4	½x3½x15	8x12	8x8	8x8	8x8	8x8	8x8
	2x6	2x4	50	45	40	0	0	0	0	0	0	"	"	"	"	"	"	"	½x4x19	10x16	10x8	"	"	"	"
	2x6	2x6	65	61	59	28	25	23	0	0	0	"	"	"	"	"	"	"	½x4x26	10x20	10x10	"	10x8	"	"
	2x8	2x6	77	73	71	33	31	29	0	0	0	2x4	2x4	2x4	2x4	4&4	4&4	2x4	2-½x6x15	12x20	12x10	8x10	10x10	8x8	8x8
	2x10	4+4	100+	100+	100+	46	44	36	0	13	0	"	"	2x6	"	"	"	"	2-½x6x21	16x20	16x12	8x12	14x12	8x10	10x8
	2x12	4+6	-	-	-	58	56	53	29	26	18	"	"	"	4&4	"	"	"	2-½x6x23	18x20	18x12	10x10	16x14	"	12x8
	2x12	6+6	-	-	-	65	59	57	32	28	26	"	"	"	"	"	"	"	2-½x6x28	20x20	20x12	10x12	18x14	8x12	14x10

58' SPAN, 6-WEB

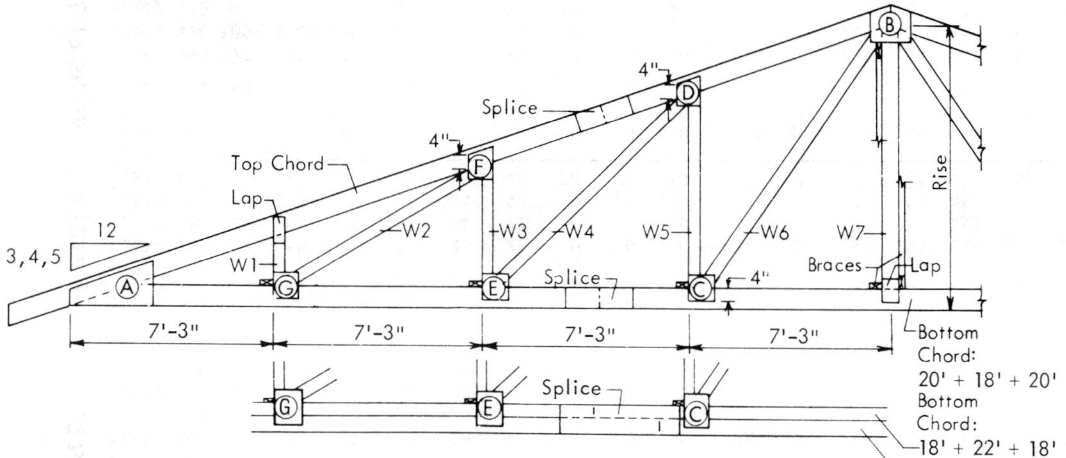

Gussets B, C, D, E, F, and G are $^3/_8$" thick plywood.

4+4, 4+6, 6+6 indicates stacked lower chord.
4&4, 6&4, indicate double web; a 2x4 is attached to the web member to increase its stiffness.

Before selecting heel gusset A, see pages 8 and 9 .

2 web lengths indicate double member, see page 10.

Web Lengths

Roof Slope	Rise	Top Chord	W1	W2	W3	W4	W5	W6	W7
3/12	7'-3"	20'+11"	2'	8'	4'	9'+8'	6'	10'+9'	7'
4/12	9'-8"	20'+11'	3'	9'	5'	10'+9'	7'+6'	12'+11'	10'
5/12	12'-1"	20'+12'	3'	9'	6'+5'	12'+11'	9'+8'	14'+12'	12'+11'

1100f Lumber

			Truss spacing, ft. — Ceiling dead load, psf — Max. snow + roof dead load, psf									Web member sizes							Gusset Sizes, in.						
			2'			4'			8'										A	B	C	D	E	F	G
Slope	Top chord	Bottom chord	0	5	8	0	5	8	0	5	8	W1	W2	W3	W4	W5	W6	W7	T H W	H W	H W	H W	H W	H W	H W
3/12 Slope	2x4	2x4	16	13	0	0	0	0	0	0	0	2x4	2x4	2x4	2x4	2x4	2x4	2x4	3/8x3½x22	8x12	8x8	8x8	8x8	8x8	8x8
	2x6	2x4	18	14	0	0	0	0	0	0	0	"	"	"	"	"	"	"	3/8x4x22	10x12	"	"	"	"	"
	2x6	2x6	27	23	20	0	0	0	0	0	0	"	"	"	"	"	"	"	3/8x4x33	"	"	"	"	"	"
	2x8	2x6	28	24	21	0	0	0	0	0	0	2x4	2x4	2x4	2x4	2x4	2x4	2x4	3/8x4x34	12x12	8x8	8x8	8x8	8x8	8x8
	2x10	4+4	39	36	34	0	12	0	0	0	0	"	"	"	"	"	"	"	½x4x28	14x16	12x10	"	10x10	"	10x8
	2x12	4+6	51	47	45	0	19	13	0	0	0	"	"	"	"	"	"	"	½x4x36	16x16	14x10	8x10	12x10	"	12x8
	2x12	6+6	58	54	52	0	21	19	0	0	0	"	"	"	"	2x6	"	"	½x4x44	16x20	16x12	8x12	16x10	"	14x8
4/12 Slope	2x4	2x4	0	16	0	0	0	0	0	0	0	2x4	2x4	2x4	2x4	2x4	2x4	2x4	3/8x3½x20	8x12	8x8	8x8	8x8	8x8	8x8
	2x6	2x4	0	19	0	0	0	0	0	0	0	"	"	"	"	"	"	"	3/8x4x21	10x12	"	"	"	"	"
	2x6	2x6	36	33	30	0	0	0	0	0	0	"	"	"	"	2x6	"	"	½x4x19	"	10x10	8x10	"	"	"
	2x8	2x6	38	35	33	0	13	0	0	0	0	2x4	2x4	2x4	2x4	2x6	2x4	2x4	½x4x19	12x16	8x10	8x10	8x8	8x8	8x8
	2x10	4+4	52	50	47	0	15	0	0	0	0	"	"	"	"	2x8	"	"	½x4x26	14x16	12x14	8x12	10x10	"	10x8
	2x12	4+6	67	63	60	29	26	20	0	0	0	"	"	"	4&4	4&4	4&4	"	½x4x32	16x16	14x10	8x10	14x10	"	12x8
	2x12	6+6	77	72	69	33	30	27	0	12	0	"	"	"	"	"	"	"	½x4x37	18x16	18x12	8x12	16x10	"	14x8
5/12 Slope	2x4	2x4	20	18	12	0	0	0	0	0	0	2x4	2x4	2x4	2x4	2x6	4&4	2x4	3/8x3½x17	8x12	8x10	8x10	8x8	8x8	8x8
	2x6	2x4	30	24	12	0	0	0	0	0	0	"	"	"	"	2x8	"	"	3/8x4x22	10x12	8x12	8x12	"	"	"
	2x6	2x6	41	39	36	0	15	0	0	0	0	"	"	"	"	4&4	"	"	½x4x18	10x16	10x8	8x8	"	"	"
	2x8	2x6	47	45	43	0	17	0	0	0	0	2x4	2x4	2x4	2x4	4&4	4&4	2x4	½x4x20	12x16	10x8	8x8	8x8	8x8	8x8
	2x10	4+4	65	62	58	28	22	0	0	0	0	"	"	2x6	4&4	"	"	"	½x4x26	14x20	12x10	"	12x10	8x10	10x8
	2x12	4+6	82	78	75	36	33	26	0	0	0	"	"	"	"	6&4	"	"	½x4x32	18x16	16x12	8x12	14x12	"	12x8
	2x12	6+6	96	89	89	42	37	35	0	16	0	"	"	2x8	"	"	"	"	½x4x38	18x20	18x12	8x14	18x14	8x12	14x8

1400f Lumber

Slope	Top chord	Bottom chord	Truss spacing 2', ceiling dead load 0 psf	2', 5	2', 8	4', 0	4', 5	4', 8	8', 0	8', 5	8', 8	W1	W2	W3	W4	W5	W6	W7	A T H W	B H W	C H W	D H W	E H W	F H W	G H W
			---Max. snow + roof dead load, psf---									Web member sizes							Gusset Sizes, in.						
3/12 Slope	2x4	2x4	0	16	14	0	0	0	0	0	0	2x4	2x4	2x4	2x4	2x4	2x4	2x4	3/8x3½x27	8x12	8x8	8x8	8x8	8x8	8x8
	2x6	2x4	0	21	17	0	0	0	0	0	0	"	"	"	"	"	"	"	½x4x16	10x12	"	"	"	"	"
	2x6	2x6	35	31	29	0	0	0	0	0	0	"	"	"	"	"	"	"	½x4x24	10x16	8x10	"	"	"	"
	2x8	2x6	37	33	31	0	12	0	0	0	0	2x4	2x4	2x4	2x4	2x4	2x4	2x4	½x4x25	12x16	10x10	8x8	8x8	8x8	8x8
	2x10	4+4	52	49	47	0	19	0	0	0	0	"	"	"	"	"	"	"	½x4x33	14x20	12x10	8x10	12x10	"	10x8
	2x12	4+6	67	62	60	0	25	23	0	0	0	"	"	"	"	2x6	"	"	½x4x40	16x20	14x14	8x12	14x10	"	12x8
	2x12	6+6	73	67	64	32	27	24	0	0	0	"	"	"	"	"	4&4	"	½x4x45	18x16	16x14	10x12	16x10	"	14x8
4/12 Slope	2x4	2x4	0	20	19	0	0	0	0	0	0	2x4	2x4	2x4	2x4	2x4	2x4	2x4	½x3½x14	8x12	8x8	8x8	8x8	8x8	8x8
	2x6	2x4	0	28	23	0	0	0	0	0	0	"	"	"	"	"	"	"	½x4x16	10x12	"	"	"	"	"
	2x6	2x6	44	41	39	0	16	13	0	0	0	"	"	"	"	2x6	"	"	½x4x23	10x16	10x12	8x10	8x10	"	"
	2x8	2x6	50	47	45	0	18	13	0	0	0	2x4	2x4	2x4	2x4	2x6	2x4	2x4	½x4x25	12x16	10x12	8x10	8x10	8x8	8x8
	2x10	4+4	69	65	62	0	27	15	0	0	0	"	"	"	"	2x8	"	"	½x4x33	16x16	14x16	8x14	12x10	"	10x8
	2x12	4+6	87	82	79	38	35	33	0	13	0	"	"	"	"	4&4	4&4	"	½x4x40	18x16	16x12	8x12	14x10	"	12x8
	2x12	6+6	96	88	89	42	37	35	0	16	12	"	"	"	"	"	"	"	½x4x46	18x20	20x12	10x10	18x12	"	14x10
5/12 Slope	2x4	2x4	26	23	21	0	0	0	0	0	0	2x4	2x4	2x4	2x4	4&4	4&4	2x4	3/8x3½x22	8x12	8x8	8x8	8x8	8x8	8x8
	2x6	2x4	40	34	29	0	0	0	0	0	0	"	"	"	"	"	"	"	½x4x16	10x12	"	"	"	"	"
	2x6	2x6	50	47	45	0	19	17	0	0	0	"	"	"	"	"	"	"	½x4x22	10x16	10x8	"	10x8	"	"
	2x8	2x6	62	59	57	0	24	18	0	0	0	2x4	2x4	2x4	2x4	4&4	4&4	2x4	½x4x25	12x20	10x10	8x8	8x8	8x8	8x8
	2x10	4+4	85	82	81	37	34	20	0	0	0	"	"	2x6	4&4	"	"	"	2-½x6x18	16x16	14x10	8x10	14x12	8x10	10x8
	2x12	4+6	100+	100+	100+	47	45	43	0	17	0	"	"	"	"	6&4	4&4	"	2-½x6x22	18x20	18x14	8x14	16x12	"	12x8
	2x12	6+6	-	-	-	51	46	44	0	21	17	"	"	2x8	"	"	"	"	2-½x6x28	"	20x14	"	18x14	8x12	14x8

1600f Lumber

Slope	Top chord	Bottom chord	Truss spacing 2', ceiling dead load 0 psf	2', 5	2', 8	4', 0	4', 5	4', 8	8', 0	8', 5	8', 8	W1	W2	W3	W4	W5	W6	W7	A T H W	B H W	C H W	D H W	E H W	F H W	G H W
			---Max. snow + roof dead load, psf---									Web member sizes							Gusset Sizes, in.						
3/12 Slope	2x4	2x4	0	21	19	0	0	0	0	0	0	2x4	2x4	2x4	2x4	2x4	2x4	2x4	½x3½x19	8x12	8x8	8x8	8x8	8x8	8x8
	2x6	2x4	0	26	22	0	0	0	0	0	0	"	"	"	"	"	"	"	½x4x19	10x12	8x10	"	"	"	"
	2x6	2x6	42	39	36	0	14	0	0	0	0	"	"	"	"	"	"	"	½x4x29	10x16	10x10	"	8x10	"	"
	2x8	2x6	44	41	38	0	15	0	0	0	0	2x4	2x4	2x4	2x4	2x4	2x4	2x4	½x4x29	12x16	10x10	8x10	8x10	8x8	8x8
	2x10	4+4	62	59	57	0	24	0	0	0	0	"	"	"	"	"	"	"	½x4x39	14x20	12x10	8x10	12x10	"	10x8
	2x12	6+4	80	75	72	35	31	29	0	13	0	"	"	"	"	"	"	"	½x4x47	18x20	16x12	10x10	14x12	"	12x8
	2x12	6+6	90	82	78	39	35	32	0	15	12	"	"	"	"	2x6	"	"	½x4x55	18x20	18x14	10x14	18x12	"	14x8
4/12 Slope	2x4	2x4	28	25	24	0	0	0	0	0	0	2x4	2x4	2x4	2x4	2x4	2x4	2x4	½x3½x16	8x12	8x8	8x8	8x8	8x8	8x8
	2x6	2x4	38	34	30	0	0	0	0	0	0	"	"	"	"	"	"	"	½x4x19	10x12	10x8	"	"	"	"
	2x6	2x6	55	51	49	23	20	18	0	0	0	"	"	"	"	2x6	4&4	"	½x4x29	10x16	10x12	8x10	8x10	"	"
	2x8	2x6	59	56	54	26	23	20	0	0	0	2x4	2x4	2x4	2x4	2x6	4&4	2x4	½x4x29	12x16	10x12	8x12	10x10	8x8	8x8
	2x10	4+4	82	79	75	36	33	25	0	0	0	"	"	"	"	2x8	"	"	2-½x6x22	16x16	14x16	8x16	12x10	"	10x8
	2x12	6+4	100+	99	95	45	43	40	0	19	0	"	"	"	"	4&4	"	"	2-½x6x28	18x20	18x12	10x12	16x12	"	12x10
	2x12	6+6	-	-	100+	51	47	45	25	21	18	"	"	"	4&4	"	"	"	2-½x6x34	20x20	20x12	"	18x12	8x10	16x10
5/12 Slope	2x4	2x4	31	28	27	0	0	0	0	0	0	2x4	2x4	2x4	2x4	4&4	4&4	2x4	½x3½x15	8x12	8x8	8x8	8x8	8x8	8x8
	2x6	2x4	48	42	38	0	0	0	0	0	0	"	"	"	"	"	"	"	½x4x19	10x16	10x8	"	"	"	"
	2x6	2x6	61	58	56	27	24	22	0	0	0	"	"	"	"	"	"	"	½x4x26	10x20	10x10	"	10x8	"	"
	2x8	2x6	75	71	69	32	30	27	0	0	0	2x4	2x4	2x4	2x4	4&4	4&4	2x4	2-½x6x15	12x20	12x10	8x10	10x10	8x8	8x8
	2x10	4+4	100+	98	99	44	43	33	0	0	0	"	"	2x6	"	"	"	"	2-½x6x21	16x20	16x12	8x12	14x12	8x10	10x8
	2x12	6+4	-	-	-	56	54	52	28	24	12	"	"	"	4&4	6&4	"	"	2-½x6x25	20x20	18x14	10x14	16x14	8x12	12x10
	2x12	6+6	-	-	-	62	57	55	31	27	24	"	"	"	"	"	"	"	2-½x6x28	20x20	20x14	"	18x14	"	14x10

60' SPAN, 6-WEB

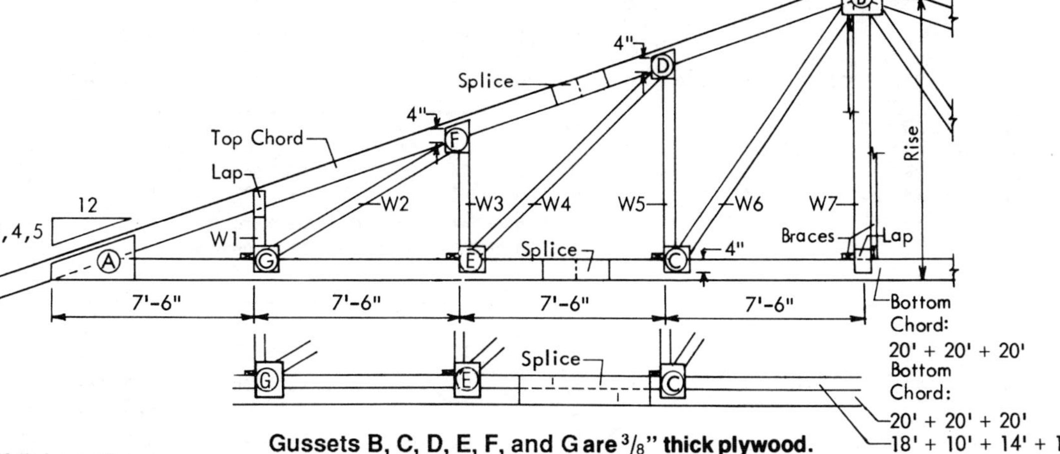

Gussets B, C, D, E, F, and G are 3/8" thick plywood.

4+4, 4+6, 6+6 indicates stacked lower chord.
4&4, 6&4, indicate double web; a 2x4 is attached to the web member to increase its stiffness.

Before selecting heel gusset A, see pages 8 and 9.

2 web lengths indicate double member, see page 10.

Web Lengths

Roof Slope	Rise	Top Chord	W1	W2	W3	W4	W5	W6	W7
3/12	7'-6"	20'+12"	2'	8'	4'	9'	6'	11'	8'
4/12	10'-0"	20'+12'	3'	9'+8'	5'	11'+10'	8'+7'	12'+11'	10'
5/12	12'-6"	20'+13'	3'	10'+9'	6'+5'	12'+11'	10'+9'	14'+13'	13'+12'

1100f Lumber

			Truss spacing, ft.									Web member sizes							Gusset Sizes, in.						
			2'			4'			8'										A	B	C	D	E	F	G
	Top chord	Bottom chord	Ceiling dead load, psf 0	5	8	0	5	8	0	5	8	W1	W2	W3	W4	W5	W6	W7	T H W	H W	H W	H W	H W	H W	H W
			---Max. snow + roof dead load, psf---																						
3/12 Slope	2x4	2x4	0	12	0	0	0	0	0	0	0	2x4	2x4	2x4	2x4	2x4	2x4	2x4	3/8x3½x21	8x12	8x8	8x8	8x8	8x8	8x8
	2x6	2x4	0	13	0	0	0	0	0	0	0	"	"	"	"	"	"	"	3/8x4x22	10x12	"	"	"	"	"
	2x6	2x6	0	22	19	0	0	0	0	0	0	"	"	"	"	"	"	"	3/8x4x33	"	"	"	"	"	"
	2x8	2x6	0	23	20	0	0	0	0	0	0	2x4	2x4	2x4	2x4	2x4	2x4	2x4	3/8x4x34	12x12	8x8	8x8	8x8	8x8	8x8
	2x10	4+4	38	35	33	0	0	0	0	0	0	"	"	"	"	"	"	"	½x4x28	14x16	12x10	"	10x10	"	10x8
	2x12	6+4	49	46	44	0	18	14	0	0	0	"	"	"	"	2x6	"	"	½x4x36	16x16	14x12	8x12	12x10	"	12x8
	2x12	6+6	56	52	50	0	21	18	0	0	0	"	"	"	"	"	"	"	½x4x44	16x20	16x12	8x12	16x10	"	14x8
4/12 Slope	2x4	2x4	0	14	0	0	0	0	0	0	0	2x4	2x4	2x4	2x4	2x4	4&4	2x4	3/8x3½x19	8x12	8x8	8x8	8x8	8x8	8x8
	2x6	2x4	0	18	0	0	0	0	0	0	0	"	"	"	"	"	"	"	3/8x4x22	10x12	"	"	"	"	"
	2x6	2x6	35	31	29	0	0	0	0	0	0	"	"	"	"	2x6	"	"	½x4x19	"	10x10	8x10	"	"	"
	2x8	2x6	37	33	31	0	12	0	0	0	0	2x4	2x4	2x4	2x4	2x6	4&4	2x4	½x4x20	12x16	10x10	8x10	8x8	8x8	8x8
	2x10	4+4	50	48	46	0	15	0	0	0	0	"	"	"	"	2x8	"	"	½x4x26	14x16	12x14	8x12	10x10	"	10x8
	2x12	6+4	64	61	58	0	25	17	0	0	0	"	"	"	"	4&4	"	"	½x4x32	16x16	14x10	8x10	14x10	"	12x8
	2x12	6+6	75	70	66	32	29	26	0	12	0	"	"	"	4&4	"	"	"	½x4x37	18x16	18x12	8x12	16x10	"	14x8
5/12 Slope	2x4	2x4	0	17	0	0	0	0	0	0	0	2x4	2x4	2x4	2x4	2x6	4&4	2x4	3/8x3½x17	8x12	8x10	8x10	8x8	8x8	8x8
	2x6	2x4	29	23	0	0	0	0	0	0	0	"	"	"	"	2x8	"	"	3/8x4x22	10x12	8x12	8x12	"	"	"
	2x6	2x6	39	36	34	0	14	0	0	0	0	"	"	"	"	4&4	"	"	½x4x18	10x16	10x8	8x8	"	"	"
	2x8	2x6	46	43	42	0	16	0	0	0	0	2x4	2x4	2x4	2x4	4&4	4&4	2x4	½x4x20	12x16	10x8	8x8	8x8	8x8	8x8
	2x10	4+4	63	60	55	0	19	0	0	0	0	"	"	2x6	"	"	"	"	½x4x26	14x20	12x10	8x8	12x10	8x10	10x8
	2x12	6+4	80	76	73	34	32	23	0	0	0	"	"	2x8	4&4	6&4	"	"	½x4x32	18x16	16x12	8x12	14x14	8x12	12x8

1400f Lumber

Slope	Top chord	Bottom chord	Truss spacing 2', Ceiling dead load 0 psf	2', 5	2', 8	4', 0	4', 5	4', 8	8', 0	8', 5	8', 8	W1	W2	W3	W4	W5	W6	W7	A T H W	B H W	C H W	D H W	E H W	F H W	G H W
			—Max. snow + roof dead load, psf—									Web member sizes							Gusset Sizes, in.						
3/12 Slope	2x4	2x4	0	15	13	0	0	0	0	0	0	2x4	2x4	2x4	2x4	2x4	2x4	2x4	3/8x3½x27	8x12	8x8	8x8	8x8	8x8	8x8
	2x6	2x4	0	19	12	0	0	0	0	0	0	"	"	"	"	"	"	"	½x4x16	10x12	"	"	"	"	"
	2x6	2x6	34	30	28	0	0	0	0	0	0	"	"	"	"	"	"	"	½x4x24	10x16	10x10	"	"	"	"
	2x8	2x6	35	32	29	0	0	0	0	0	0	2x4	2x4	2x4	2x4	2x4	2x4	2x4	½x4x25	12x16	10x10	8x8	8x8	8x8	8x8
	2x10	4+4	50	47	45	0	18	0	0	0	0	"	"	"	"	"	"	"	½x4x34	14x20	12x10	8x10	12x10	"	10x8
	2x12	6+4	64	60	58	0	24	22	0	0	0	"	"	"	"	2x6	"	"	½x4x40	16x20	14x14	8x12	14x10	"	12x8
	2x12	6+6	71	65	62	31	26	23	0	0	0	"	"	"	"	"	4&4	"	½x4x45	18x16	18x14	10x12	16x10	"	14x8
4/12 Slope	2x4	2x4	0	19	17	0	0	0	0	0	0	2x4	2x4	2x4	2x4	2x4	2x4	2x4	3/8x3½x24	8x12	8x8	8x8	8x8	8x8	8x8
	2x6	2x4	0	26	21	0	0	0	0	0	0	"	"	"	"	2x6	"	"	½x4x16	10x12	8x10	8x10	"	"	"
	2x6	2x6	42	39	37	0	15	12	0	0	0	"	"	"	"	"	"	"	½x4x23	10x16	10x12	"	"	"	"
	2x8	2x6	48	45	43	0	18	12	0	0	0	2x4	2x4	2x4	2x4	2x8	2x4	2x4	½x4x25	12x16	10x14	8x12	8x10	8x8	8x8
	2x10	4+4	66	63	60	0	25	12	0	0	0	"	"	"	"	4&4	"	"	½x4x33	16x16	14x12	8x10	12x10	"	10x8
	2x12	6+4	84	80	77	36	34	31	0	12	0	"	"	"	"	"	4&4	"	2-½x6x28	18x16	16x12	8x12	14x10	"	12x8
	2x12	6+6	93	85	81	40	36	33	0	15	0	"	"	"	"	"	"	"	½x4x46	18x20	20x12	10x10	18x12	"	14x10
5/12 Slope	2x4	2x4	24	22	20	0	0	0	0	0	0	2x4	2x4	2x4	2x4	4&4	4&4	2x4	3/8x3½x22	8x12	8x8	8x8	8x8	8x8	8x8
	2x6	2x4	39	33	28	0	0	0	0	0	0	"	"	"	"	"	"	"	½x4x16	10x12	"	"	"	"	"
	2x6	2x6	48	45	43	0	18	16	0	0	0	"	"	"	"	"	"	"	½x4x22	10x16	10x8	"	10x8	"	"
	2x8	2x6	60	57	55	0	23	16	0	0	0	2x4	2x4	2x6	2x4	4&4	4&4	2x4	½x4x25	12x20	10x10	8x8	10x10	8x10	8x8
	2x10	4+4	83	79	78	36	32	18	0	0	0	"	"	"	4&4	6&4	"	"	2-½x6x18	16x16	14x12	8x12	14x12	"	10x8
	2x12	6+4	100+	100	100	45	43	40	0	16	0	"	"	2x8	"	"	"	"	2-½x6x22	18x20	18x14	10x12	16x14	8x12	12x8
	2x12	6+6	-	-	-	49	45	43	0	22	15	"	"	"	"	"	"	"	2-½x6x28	"	20x14	8x14	18x14	8x14	14x10

1600f Lumber

Slope	Top chord	Bottom chord	Truss spacing 2', Ceiling dead load 0 psf	2', 5	2', 8	4', 0	4', 5	4', 8	8', 0	8', 5	8', 8	W1	W2	W3	W4	W5	W6	W7	A T H W	B H W	C H W	D H W	E H W	F H W	G H W
			—Max. snow + roof dead load, psf—									Web member sizes							Gusset Sizes, in.						
3/12 Slope	2x4	2x4	0	20	18	0	0	0	0	0	0	2x4	2x4	2x4	2x4	2x4	2x4	2x4	½x3½x18	8x12	8x8	8x8	8x8	8x8	8x8
	2x6	2x4	0	25	21	0	0	0	0	0	0	"	"	"	"	"	"	"	½x4x20	10x12	8x10	"	"	"	"
	2x6	2x6	41	37	34	0	13	0	0	0	0	"	"	"	"	"	"	"	½x4x29	10x16	10x10	"	8x10	"	"
	2x8	2x6	43	40	36	0	14	0	0	0	0	2x4	2x4	2x4	2x4	2x4	2x4	2x4	½x4x30	12x16	10x10	8x10	8x10	8x8	8x8
	2x10	4+4	60	57	55	0	23	14	0	0	0	"	"	"	"	"	"	"	½x4x39	14x20	12x10	"	12x10	"	10x8
	2x12	6+4	77	72	69	33	30	28	0	13	0	"	"	"	"	2x6	"	"	½x4x47	18x20	16x14	10x12	14x12	"	12x8
	2x12	6+6	87	80	76	38	33	31	0	14	0	"	"	"	"	"	"	"	½x4x56	"	18x14	10x14	18x12	"	14x8
4/12 Slope	2x4	2x4	27	24	22	0	0	0	0	0	0	2x4	2x4	2x4	2x4	2x4	2x4	2x4	½x3½x16	8x12	8x8	8x8	8x8	8x8	8x8
	2x6	2x4	37	33	28	0	0	0	0	0	0	"	"	"	"	"	4&4	"	½x4x19	10x12	10x8	"	"	"	"
	2x6	2x6	52	49	47	0	19	17	0	0	0	"	"	"	"	2x6	"	"	½x4x28	10x16	10x12	8x10	8x10	"	"
	2x8	2x6	58	54	52	0	22	19	0	0	0	2x4	2x4	2x4	2x4	2x6	4&4	2x4	½x4x30	12x16	10x12	8x12	10x10	8x8	8x8
	2x10	4+4	80	76	73	34	32	23	0	0	0	"	"	"	"	2x8	"	"	2-½x6x22	16x16	14x16	8x16	12x10	"	10x10
	2x12	6+4	100+	96	92	44	41	39	0	18	0	"	"	"	"	4&4	"	"	2-½x6x28	18x20	18x12	10x12	16x12	"	12x10
	2x12	6+6	-	100+	100+	50	46	43	0	20	17	"	"	"	"	"	"	"	2-½x6x34	20x20	20x12	"	18x12	8x10	16x10
5/12 Slope	2x4	2x4	29	27	25	0	0	0	0	0	0	2x4	2x4	2x4	2x4	4&4	4&4	2x4	½x3½x14	8x12	8x8	8x8	8x8	8x8	8x8
	2x6	2x4	47	41	36	0	0	0	0	0	0	"	"	"	"	"	"	"	½x4x19	10x16	10x8	"	"	"	"
	2x6	2x6	58	55	53	25	22	21	0	0	0	"	"	"	"	"	"	"	½x4x26	10x20	10x10	"	10x8	"	"
	2x8	2x6	72	69	67	31	29	25	0	0	0	2x4	2x4	2x4	2x4	4&4	4&4	2x4	2-½x6x15	14x16	12x10	8x10	10x10	8x8	8x8
	2x10	4+4	99	95	96	43	41	29	0	0	0	"	"	2x6	"	"	"	"	2-½x6x21	16x20	16x12	8x12	14x12	8x10	10x8
	2x12	6+4	-	100+	100+	54	52	50	0	23	0	"	"	2x8	4&4	6&4	"	"	2-½x6x26	20x20	18x14	10x14	16x16	8x14	12x10
	2x12	6+6	-	-	-	60	56	53	30	26	23	"	4&4	"	"	"	"	"	2-½x6x28	"	20x14	"	18x16	"	14x10

Design Notes

This information is intended for engineers to modify designs with their own judgment, to modify MWPS plans for special conditions, and to compare with code or other requirements.

Three major design standards are incorporated:

- *National Design Specification for Wood Construction,* 1977 Edition, National Forest Products Association. 1980 Design Values Supplement. (Abbr: NDS)
- *American National Standard Building Code Requirements for Minimum Design Loads in Buildings and Other Structures,* American National Standards Institute standard A58.1-1972 (Abbr: ANSI)
- *Plywood Design Specification,* 1974, American Plywood Association (Abbr: PDS)

This handbook is listed in *US Housing and Urban Development Handbook 5950.2 REV-1* in its section on "Other Acceptable Criteria".

Loads

Snow

50-yr snow loads in Fig 4 are from ASAE.

Trusses were designed for uniform load on the horizontal projection. Load on only one slope was not checked.

Dead

Designs are provided for ceiling dead loads of 0, 5, and 8 psf. Add truss and roof dead loads to snow load to get the design load to be found in the design tables.

Wind

Under most wind loads, the tension chords become compression members, requiring additional stiffeners, and stress reverses in the tension members, sometimes requiring additional member area. Webs were assumed not laterally supported.

ANSI wind velocities are from analyses of the fastest mile per year at a standard elevation of 30'. The wind pressures include gusts, but are not reduced for buildings less than 30' high.

ANSI wind speeds of 80 mph for a 25-year recurrence interval include all of the U.S. except SE and

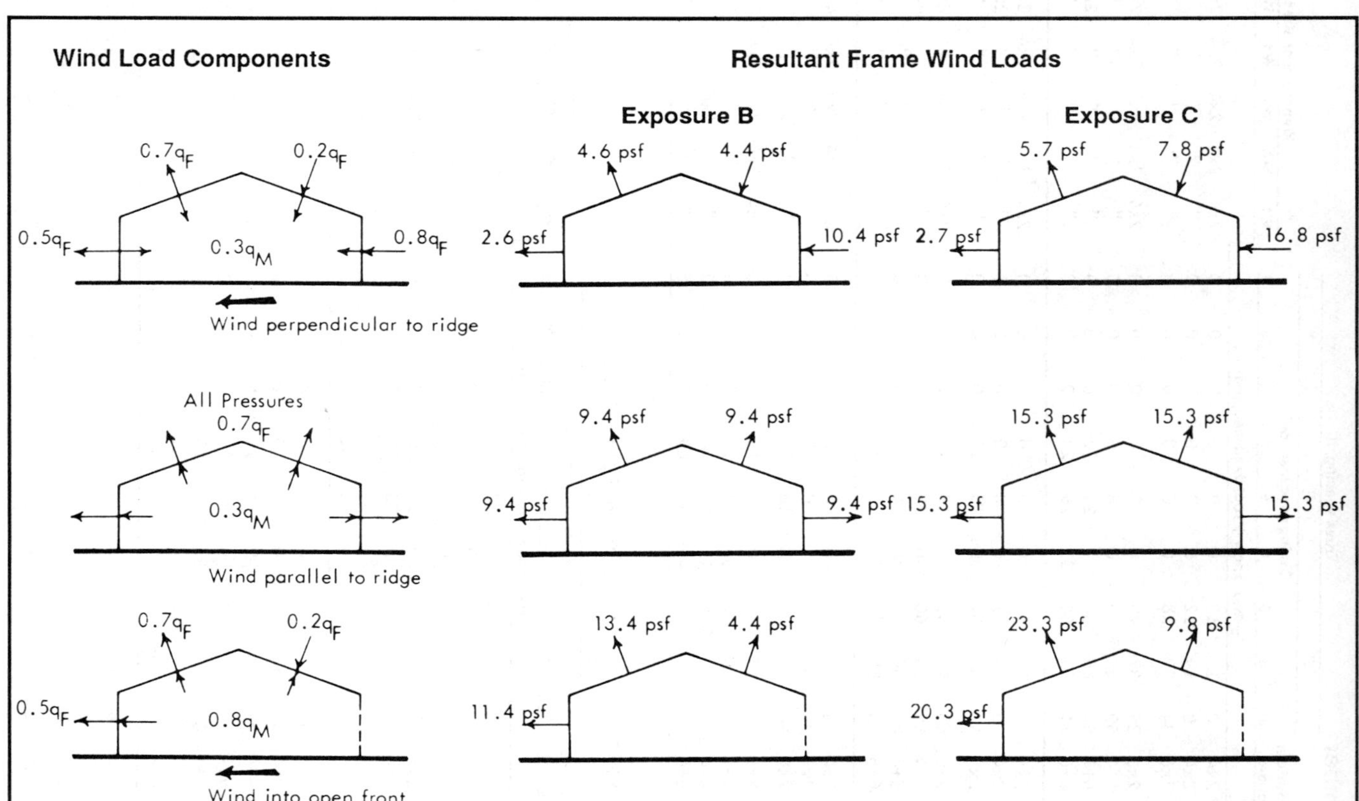

Fig 22. 80mph wind loads on building frames.
Exposure B: qm = 8 psf internal pressure; qf = 10 psf external pressure.
Exposure C: qm = 16 psf inernal pressure; qf = 15 psf external pressure.

NW coastal areas. ANSI pressures for Exposure B (suburban areas, towns, city outskirts, wooded areas, and rolling terrain) and qf for building frames are not increased for tributary areas less than 1000 sq ft. For flat coastal belts, use ANSI Exposure C wind pressures to check MWPS designs. See Fig 22.

All trusses were checked against ANSI exposure B with wind parallel and perpendicular to the building using a statically determinate pinned-joint computer program. Maximum L/d for webs under wind stress was increased to 80.

Materials and Stress Selection

Adjustments for Load Duration and Repetitive Members

NDS and standard practice permit increasing wood design stresses for short load duration, e.g. 15% for snow and 33% for wind. Commonly, the stress increase is applied to the combined load. Also, allowable stresses are increased 15% for members spaced no more than 24" apart.

The 1986 NDS applies duration-of-load factors to short and intermediate columns but not to long ones. The factors are applied to Fc, but not to Fc′; Fc is not in the long-column equation for Fc′. This is a change from previous editions.

The upper chords in these trusses are not long columns because roof purlins are assumed to provide adequate lateral bracing. Some longer compression webs are marginally long columns under gravity loads. Some tension webs are long columns under wind stress reversal.

Lumber

Because strength is a function of moisture content, and member selection depends on both the size of the member and its strength, establish three elements before designing in wood: moisture content during use, size during use, and quality or allowable stresses.

Moisture content of the wood during use, as well as at a time of milling, is of interest. Strength of the lumber, strength and perhaps durability of glue joints, and performance of nails and other hardware fasteners, are all affected.

Fig 23 relates relative humidity, dry bulb temperature, and equilibrium moisture content of wood. Even in livestock buildings, it is unlikely that relative humidity is high enough long enough to raise the moisture content of lumber appreciably above 15%. Lumber shrinks during drying, so NDS stresses apply to moisture content and size at time of surfacing. NDS stresses still apply if the lumber is further dried, even though it shrinks.

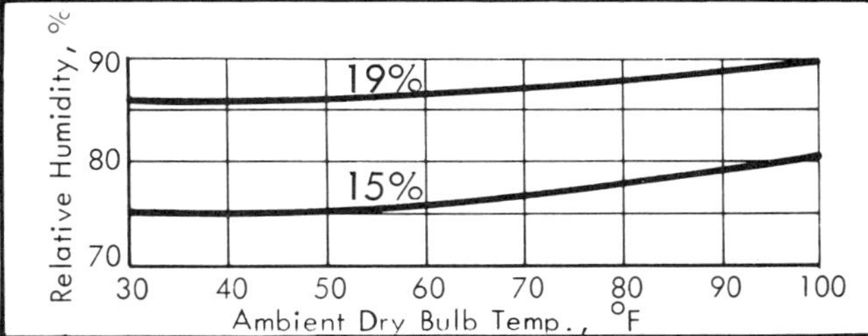

Fig 23. Equilibrium moisture content of wood.
Wood Handbook, No. 72, USDA.

Because of the low tensile strength of 2x8s, 2x10s, and 2x12s, these member sizes are not used in lower chords. All lower chords requiring a member larger than a 2x6 are made from 2x4s and 2x6s stacked vertically. Note: in member 4+6, the 2x4 is above the 2x6, which is better for ceiling loads. See MSR Lumber.

Properties of the lumber used in the designs are listed in Table 9. The bending properties, I and S, for stacked bottom chords assume the two members act independently, which is conservative. The equivalent b and d are the dimensions given the computer design program.

For 4+6, the b and d were calculated to give the combined member the same A and S as the sum of the 2x4 and 2x6. The tabulated I was calculated from the equivalent b and d; I of 2x4 + 2x6 is 25.062.

Table 9. Lumber dimension properties.

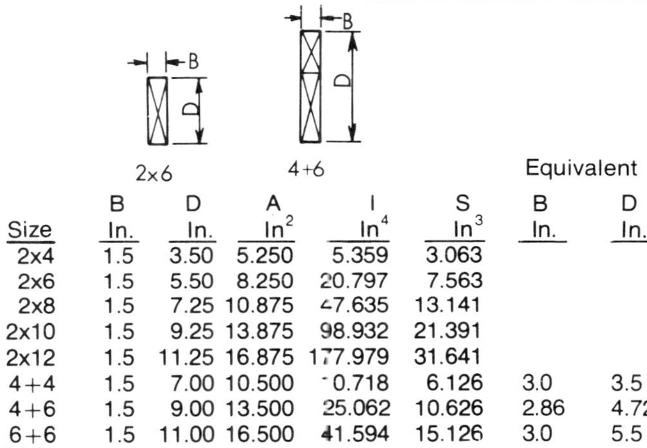

Size	B In.	D In.	A In^2	I In^4	S In^3	Equivalent B In.	Equivalent D In.
2x4	1.5	3.50	5.250	5.359	3.063		
2x6	1.5	5.50	8.250	20.797	7.563		
2x8	1.5	7.25	10.875	47.635	13.141		
2x10	1.5	9.25	13.875	98.932	21.391		
2x12	1.5	11.25	16.875	177.979	31.641		
4+4	1.5	7.00	10.500	10.718	6.126	3.0	3.5
4+6	1.5	9.00	13.500	25.062	10.626	2.86	4.72
6+6	1.5	11.00	16.500	41.594	15.126	3.0	5.5

Plywood Gussets

Plywood stresses and shear thicknesses are from the 1974 Plywood Design Specification.

Table 10. Effective thickness, sheer-through-plies.

Plywood thickness	No. of plies	Identification index	Shear thickness
3/8"	3	24/0	0.293"
1/2"	5	32/16	0.529"
5/8"	5	42/20	0.553"
3/4"	5	48/24	0.573"
2-1/2" sheets		32/16	1.058"

Analysis

The Purdue Plane Structures Analyzer (PPSA II, 1979) is a FORTRAN computer program developed by Suddarth and others to analyze 2-dimensional wood structures. Especially since its 1979 revision, PPSA seems to be becoming a standard of engineering practice in wood truss design. The MWPS decided that its use was justified in a major revision of this handbook. PPSA's stiffness matrix method allows a wide variety of structural configurations. It directly accounts for secondary stresses in trusses with continuous members or fixed joints and avoids some of the simplifying assumptions of pinned-joint analysis and moment coefficients. PPSA II includes 1977 NDS provisions.

Why PPSA?

The rigid heel joint of a truss under gravity loads rotates and introduces secondary stresses in the bottom chord.

Heel joint rotation per pound of applied load depends on truss configuration, stiffness of the various members, position of the support under the joint, bending in the first panel of the top chord, etc. Moment transfer to the bottom chord depends on the relative stiffness of the two chord members, truss configuration, etc. Therefore, although a pinned-joint program can approximate the introduced moment by adding a moment to the axial stresses, a fixed coefficient cannot be accurate.

The MWPS could not find a relationship to add to a pinned-joint program to approximate fixed-end conditions more nearly than a single moment coefficient. Various relationships for top and bottom chord properties and panel lengths were tried, but did not predict secondary stress effects within tolerable limits.

Fixed-joint and pinned-joint designs were compared for a few cases in 36' to 50' spans. Allowable loads for the indeterminate analysis were higher for trusses of relatively slender members and lower for trusses of relatively stiff members. If generally true, these results suggest that lighter trusses are somewhat stronger than we had been predicting and that larger and heavier trusses have been somewhat underdesigned.

Structural Analog

PPSA II was used for gravity load design of the trusses in this book. See Loads, Wind.

To use the program, all framed structures are first reduced to line-drawing models (analogs). Fig 24 & 25.

In selecting the heel joint analog, it was found that design is fairly sensitive to the location of the support and fictitious member C. Designs are not very sensitive to the properties selected for the fictitious members as long as they are considerably stiffer than a pinned joint.

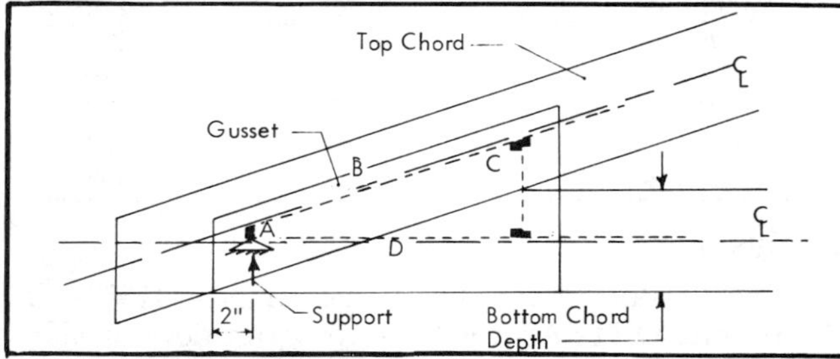

Fig 24. Heel joint analog.
Dotted lines indicate the analog PPSA analysis. Members A to D were fictitious; they were rigidly connected and had 10 times the MOE of the lumber. Support was 2" from the outside edge of the gusset for all trusses.

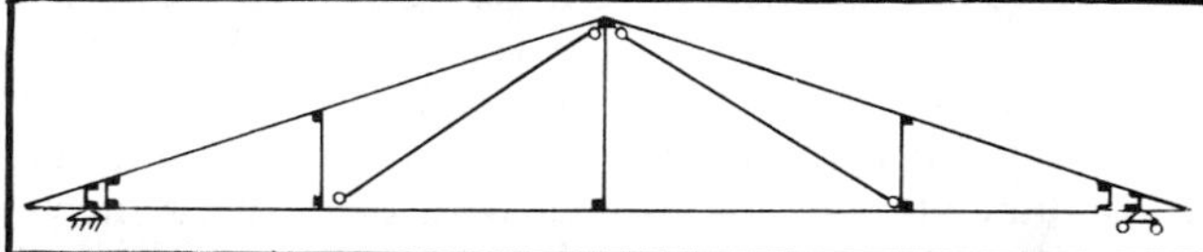

Fig 25. Web and ridge joint analogs.
All members were rigidly connected to joints except diagonal webs, which were pinned.

Coordinates for all joints and members were measured along the centerlines. PPSA reduces the centerline length of compressive members to allow for cut lengths of members and for joints.

Assumptions: Webs are not laterally braced. Top chords are laterally supported 24" o.c. by roofing. Bottom chords are braced continuously where a ceiling load occurs, and are braced at the panel points with no ceiling load.

Interaction

Interaction refers to the dual effect of bending and axial stresses. It is also used here for cases of axial only or bending only, which are simply special cases of the interaction equation with one term zero. The interaction number measures the proportion of a member's strength used up by the applied loads: if less than one, strength remains; if more than one, the member is overloaded.

For gravity loads the slenderness ratio, L/D, for a compression member is limited to 50, and for a tension member to 80. PPSA reduces the effective in-plane length of members in buckling due to gravity loads by 0.8 for chords and 0.9 for webs.

The allowable buckling stress, F'c, is no greater than Fc:

- Short Column, L/D = 11 or less:
  $F_c' = F_c$
- Intermediate column, L/D between 11 and K:
  $K = 0.671\,(E/F_c)^{0.5}$
  $F_c' = F_c(1-(L/KD)^4/3)$
- Long column, L/D = K or greater:
  $F_c' = 0.30E/(L/D)^2$

The interaction equations for combined axial and bending loads are:

- Tension + bending:

$$\frac{P/A}{F_t} + \frac{M/S}{F_b} \leq 1.0$$

- Compression + bending:

$$\frac{P/A}{F_c'} + \frac{M/S}{F_b - J(P/A)} \leq 1.0$$

$$J = \frac{(L/D) - 11}{K - 11}$$

$$0 \leq J \leq 1.0$$

Where: P=axial force; M=bending moment; A=area of member's cross section; S=section modulus; E=modulus of elasticity; D=least width; Ft=allowable tensile stress; Fc= allowable compressive stress on a short column; F'c=allowable compressive stress on an intermediate or long column; Fb=allowable stress.

The interaction is calculated at each end of a compression member using the short-column allowable stress. A third interaction is calculated at the point of maximum moment using the compressive stress at that point and the effective column length. The highest of the three interactions governs.

PPSA checks stress in the member both parallel and perpendicular to plane and reports the critical stress. Loads are reported based on full stress in either the top or bottom chord.

Modifications to PPSA II

PPSA is arranged to allow several different loading conditions for one structural configuration. The program was modified to loop through the many truss configurations in this book, redefine point locations, slopes, member sizes, spacing, etc., and to print the results. Fig 26.

Because PPSA is a checking (indeterminate) program, member sizes must first be estimated and then checked for adequacy. A 50 psf roof load was applied to a given truss and interaction values calculated. The 50 psf roof load was then divided by the highest interaction value ("Big") for the chord members. This new load was used to recalculate interactions. For example, in Fig 26 the largest chord interaction is 1.667 under the 50 psf load. New interactions were then calculated based on a 50/1.667 or 30 psf roof load. Iteration continued until the largest chord interaction value fell between 0.95 and 1.02. The load for the last iteration is the design roof load. Note that interaction is not linear with load because of the "J(P/A)" factor in the denominator of the NDS compression-plus-bending formula.

All webs were assumed to be 2x4s without lateral bracing. Web sizes were later adjusted based on interaction as shown in Table 12. All webs were then checked for wind and further modified as needed.

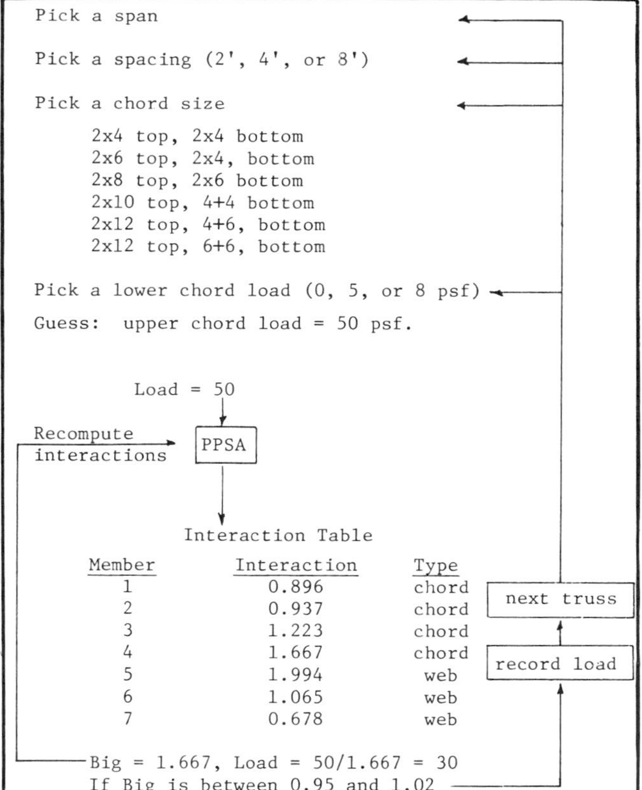

Fig 26. Modified PPSA II for MWPS truss designs.

Table 12. Adjusting web sizes from 2x4 interactions.

If 2x4 Interaction is:	Then Web is:
1.00 or less	2x4 OK
1.01 - 1.57	2x6
1.58 - 2.07	2x8
between 2.08 and table value listed below	2x4 web + 2x4 stiffener
Higher than table value	2x6 web + 2x4 stiffener

Up to 2.08, adjustments are based on the ratio of member areas. (1.57 = 2x6 Area ÷ 2x4 area) Table values are based on halving L/D and the corresponding change from long to intermediate column.

L/D	1600f	1400f and 1100f
30	2.10	2.14
32	2.36	2.40
34	2.61	2.67
36	2.87	2.93
38	3.11	3.41
40	3.34	3.62
42	3.55	3.79
44	3.72	3.92
46	3.84	3.99
48	3.91	3.99
50	3.91	3.99

Gusset Design and Selection

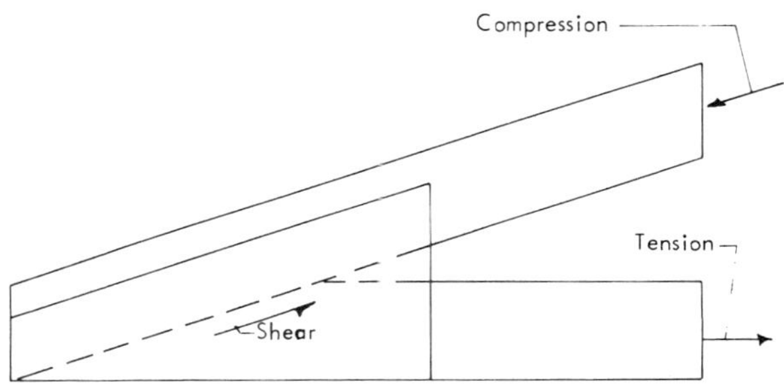

Heel Gussets

Gusset length and thickness prevent shear failure (250 psi) in the plywood between the chords. The heel glue area prevents rolling shear failure (53 psi) between the plywood and the chords. Both allowable shear values increase with angle to face grain, so the joints have reserve strength to carry transverse chord loads and moments. Place heel joint plywood with the face grain horizontal.

The heel gusset must at least cover the scarf cut, which depends on member size and roof slope.

Table 11. Scarf cut lengths.

Roof slope	Lower chord size 2x4	2x6	4+4	4+6	6+6
	---Scarf cut, inches---				
3/12	14"	22"	28"	36"	44"
4/12	10½	16½	21	27	33
5/12	8½	13¼	17	21¾	26½

The maximum length of heel gusset before going to the next larger thickness is arbitrary. Several factors apply.

- All gussets except the heel are 3/8", so 3/8" is used where reasonable, because: the trusses stack better, only one plywood thickness need be purchased, and 3/8" thickness with the correct number of plies is more readily available in some areas than the 5-ply 1/2".
- When the heel gusset is longer than the scarf cut, the plywood between the upper face of the bottom chord and the lower face of the top chord is unsupported and tends to buckle as the top chord bends.
- The longer the heel gusset, the more "waste," because the additional height does not contribute to shear strength.

  Heel gussets were limited to the length of the scarf joint plus about 14".

Minimum heel gusset height is measured at the outer end of the truss. The gusset extends from the bottom of the bottom chord to 4" up on the upper chord, but 6" up on the upper chord with gussets that are two layers of 1/2" plywood.

If the lower chord is offset, increase the depth of the gusset the amount of the offset and use the tabulated gusset length, Fig 7b.

Ridge and Web Gussets

These gussets were selected for enough glue area to carry member tension loads into the joints. The limiting criterion is rolling shear (53 psi).

Ridge gussets cover the full depth of the top chord. Web gussets cover top chords by 4" and the entire width of vertical webs. Gussets cover the full depth of 2x4 bottom chords, and all but 1" for other bottom chord sizes. Tabulated gussets include 2" additional width for 2x6 vertical webs.

Laps and Splices

Lap joints connect the shortest web to the top chord and the kingpost to the bottom chord. They carry small loads and are not designed for each truss.

Splices were designed to carry the full strength of the members connected. For illustrations of laps and splices see Figs 8-10.

Summary Page 2-web

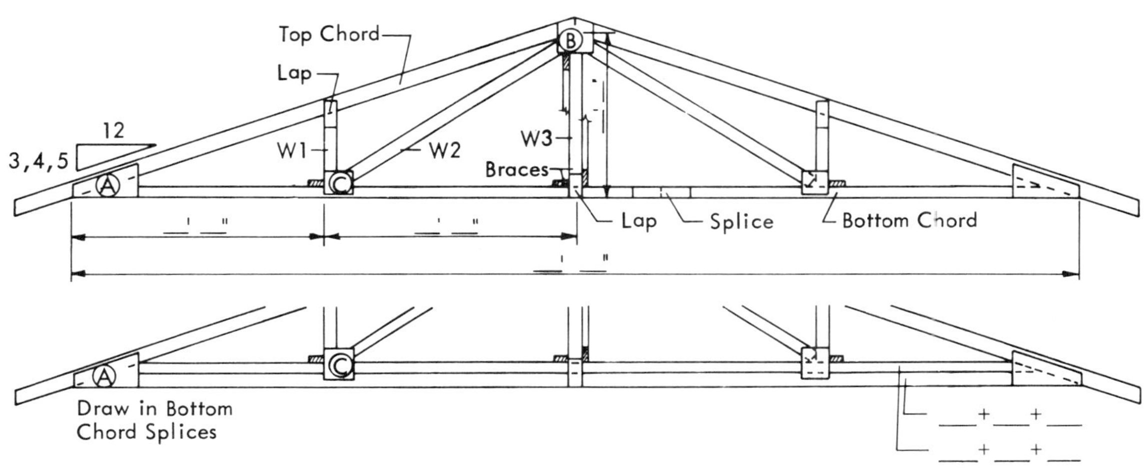

BOARD	DESCRIPTION	NO./ TRUSS	TOTAL/ BUILDING
Top Chord	2 x __ x __ '	2	____
Bottom Chord	2 x __ x __ '	__	____
	2 x __ x __ '	__	____
	2 x __ x __ '	__	____
	2 x __ x __ '	__	____
W1	2 x 4 x __ '	2	____
W2	2 x __ x __ '	2	____
	2 x __ x __ '	2	____
W3	2 x __ x __ '	1	____

Span & Web __________

Lumber Group __________

Roof Slope __________

Spacing __________

Ceiling Dead Load __________

Roof Dead Load __________

Snow Load __________

Total Snow + roof Dead

Dimensions from Page __________

No. Trusses needed ________

JOINTS -- GUSSETS

JOINT	GUSSET SIZE, inches Thickness by Height by Width	NO./ TRUSS	TOTAL/ BUILDING
*A	____ x ____ x ____	4	____
B	3/8" x ____ x ____	2	____
C	3/8" x ____ x ____	4	____

*Check Heel Gussets, page 8, before final height (H) selection

Lap Joints**	____ x ____ x ____	6	____
Splice Joints**	____ x ____ x ____	____	____

**See page 9

GUSSETS T = Thickness

H
W
Joint A

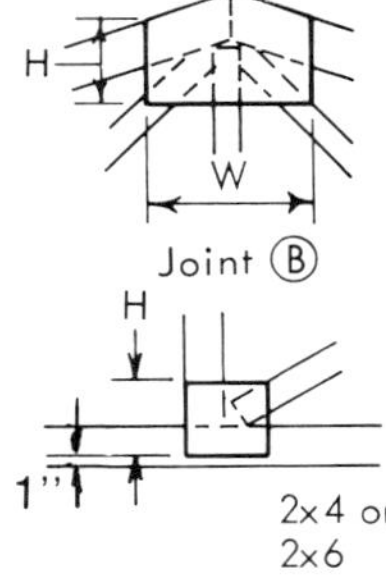

Joints C, E, G Single Chord
C Stacked Chord

Wind Anchorage The required number of bolts or framing anchors is listed on page 13.

Use ____ Anchors or ____ Bolts for each truss. TOTAL/BUILDING ______.

Wind Bracing Read pages 12 and 13 carefully. The bracing described is required for safety both during and after truss erection.

2x4 Bottom Chord Bracing 3 times building length = TOTAL lineal feet required _____ft.

These braces must be the full length of the building. They are not needed in buildings that will have a rigid ceiling.

Cross Bracing 2 x __ x __ ft. TOTAL lineal feet required ____.

Knee Bracing 2 x __ x __ ft. TOTAL lineal feet required ____.

Roofing Material __________ Consult your lumber dealer. Read pages 14 and 15 for the support needed.

Ceiling Material __________ The material manufacturer will usually specify the framing to use.

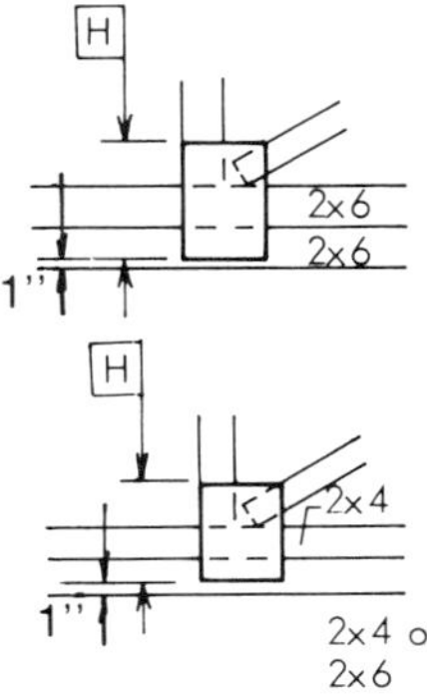

Joints C, E, G Stacked Chord

Summary Page 2-web

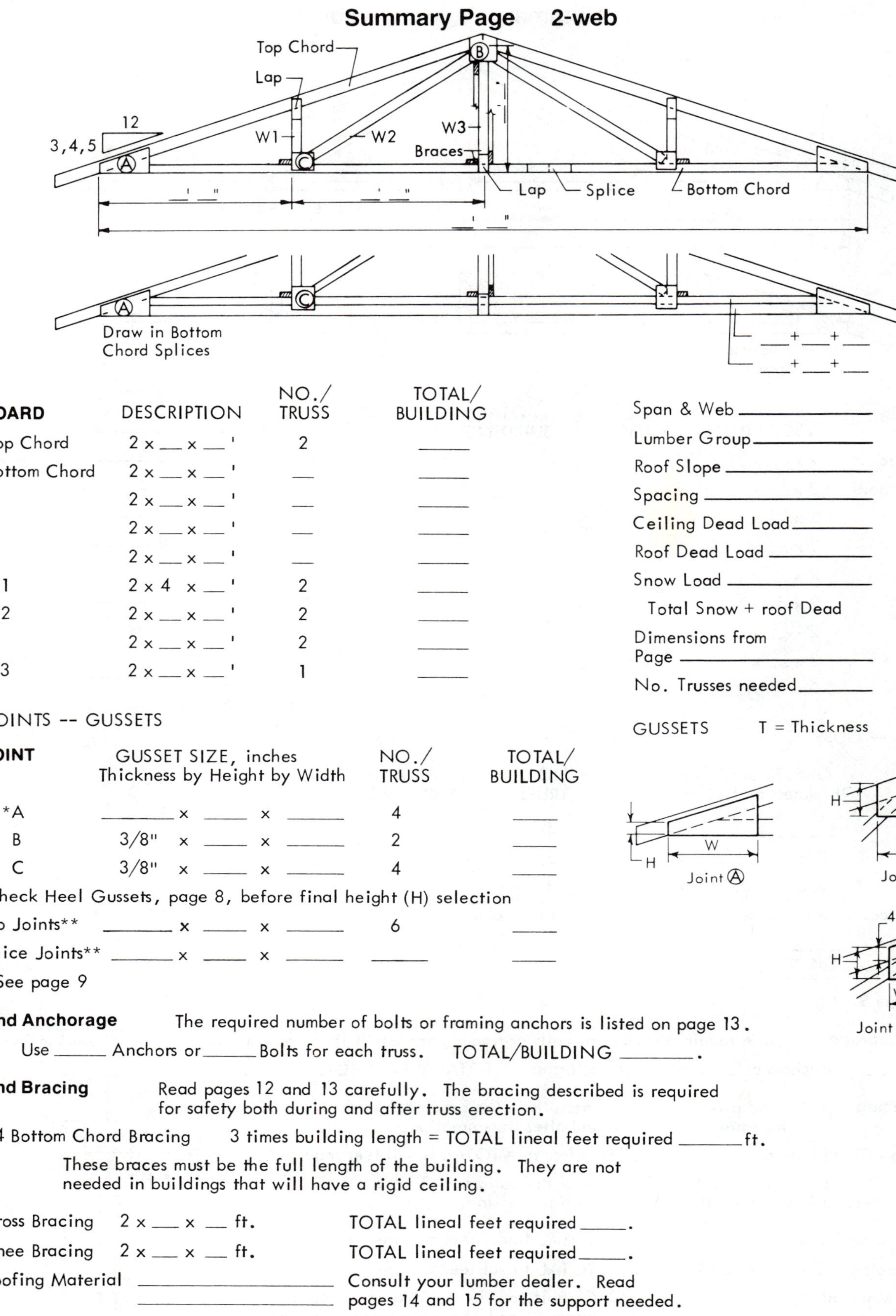

BOARD	DESCRIPTION	NO./ TRUSS	TOTAL/ BUILDING
Top Chord	2 x __ x __ '	2	____
Bottom Chord	2 x __ x __ '	__	____
	2 x __ x __ '	__	____
	2 x __ x __ '	__	____
	2 x __ x __ '	__	____
W1	2 x 4 x __ '	2	____
W2	2 x __ x __ '	2	____
	2 x __ x __ '	2	____
W3	2 x __ x __ '	1	____

Span & Web ________

Lumber Group ________

Roof Slope ________

Spacing ________

Ceiling Dead Load ________

Roof Dead Load ________

Snow Load ________

Total Snow + roof Dead

Dimensions from Page ________

No. Trusses needed ________

JOINTS -- GUSSETS

JOINT	GUSSET SIZE, inches Thickness by Height by Width	NO./ TRUSS	TOTAL/ BUILDING
*A	____ x ____ x ____	4	____
B	3/8" x ____ x ____	2	____
C	3/8" x ____ x ____	4	____

*Check Heel Gussets, page 8, before final height (H) selection

Lap Joints**	____ x ____ x ____	6	____
Splice Joints**	____ x ____ x ____	____	____

**See page 9

GUSSETS T = Thickness

Wind Anchorage The required number of bolts or framing anchors is listed on page 13.

Use ____ Anchors or ____ Bolts for each truss. TOTAL/BUILDING ______.

Wind Bracing Read pages 12 and 13 carefully. The bracing described is required for safety both during and after truss erection.

2x4 Bottom Chord Bracing 3 times building length = TOTAL lineal feet required _____ft.

These braces must be the full length of the building. They are not needed in buildings that will have a rigid ceiling.

Cross Bracing 2 x __ x __ ft. TOTAL lineal feet required ____.

Knee Bracing 2 x __ x __ ft. TOTAL lineal feet required ____.

Roofing Material __________ Consult your lumber dealer. Read pages 14 and 15 for the support needed.

Ceiling Material __________ The material manufacturer will usually specify the framing to use.

Summary Page 4-web

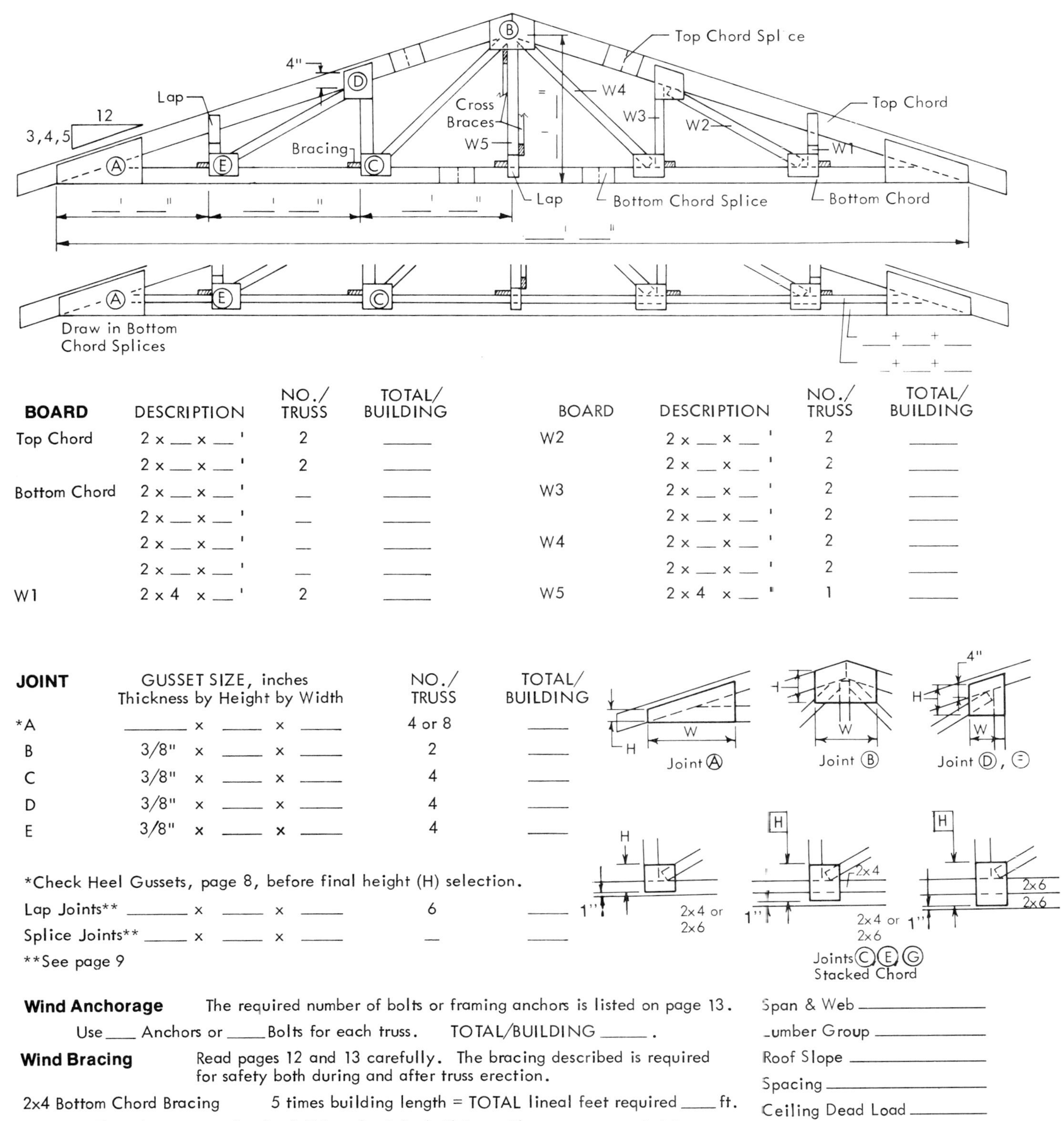

BOARD	DESCRIPTION	NO./ TRUSS	TOTAL/ BUILDING	BOARD	DESCRIPTION	NO./ TRUSS	TOTAL/ BUILDING
Top Chord	2 x ___ x ___ '	2	_____	W2	2 x ___ x ___ '	2	_____
	2 x ___ x ___ '	2	_____		2 x ___ x ___ '	2	_____
Bottom Chord	2 x ___ x ___ '	___	_____	W3	2 x ___ x ___ '	2	_____
	2 x ___ x ___ '	___	_____		2 x ___ x ___ '	2	_____
	2 x ___ x ___ '	___	_____	W4	2 x ___ x ___ '	2	_____
	2 x ___ x ___ '	___	_____		2 x ___ x ___ '	2	_____
W1	2 x 4 x ___ '	2	_____	W5	2 x 4 x ___ '	1	_____

JOINT	GUSSET SIZE, inches Thickness by Height by Width	NO./ TRUSS	TOTAL/ BUILDING
*A	_____ x _____ x _____	4 or 8	_____
B	3/8" x _____ x _____	2	_____
C	3/8" x _____ x _____	4	_____
D	3/8" x _____ x _____	4	_____
E	3/8" x _____ x _____	4	_____

*Check Heel Gussets, page 8, before final height (H) selection.

Lap Joints**	_____ x _____ x _____	6	_____
Splice Joints**	_____ x _____ x _____	___	_____

**See page 9

Wind Anchorage The required number of bolts or framing anchors is listed on page 13.

Use ___ Anchors or ____ Bolts for each truss. TOTAL/BUILDING _____ .

Wind Bracing Read pages 12 and 13 carefully. The bracing described is required for safety both during and after truss erection.

2x4 Bottom Chord Bracing 5 times building length = TOTAL lineal feet required ____ ft.

These braces must be the full length of the building. They are not needed in buildings that will have a rigid ceiling.

Cross Bracing 2 x ___ x ___ ft. TOTAL lineal feet required _____ .

Knee Bracing 2 x ___ x ___ ft. TOTAL lineal feet required _____ .

Roofing Material ____________________ Consult your lumber dealer. Read page 14 and 15 for the support needed.

Ceiling Material ____________________ The material manufacturer will usually specify the framing to use.

Span & Web __________

Lumber Group __________

Roof Slope __________

Spacing __________

Ceiling Dead Load __________

Roof Dead Load __________

Snow Load __________

Total Snow + roof Dead

Dimensions from Page __________

No. Trusses needed __________

Summary Page 4-web

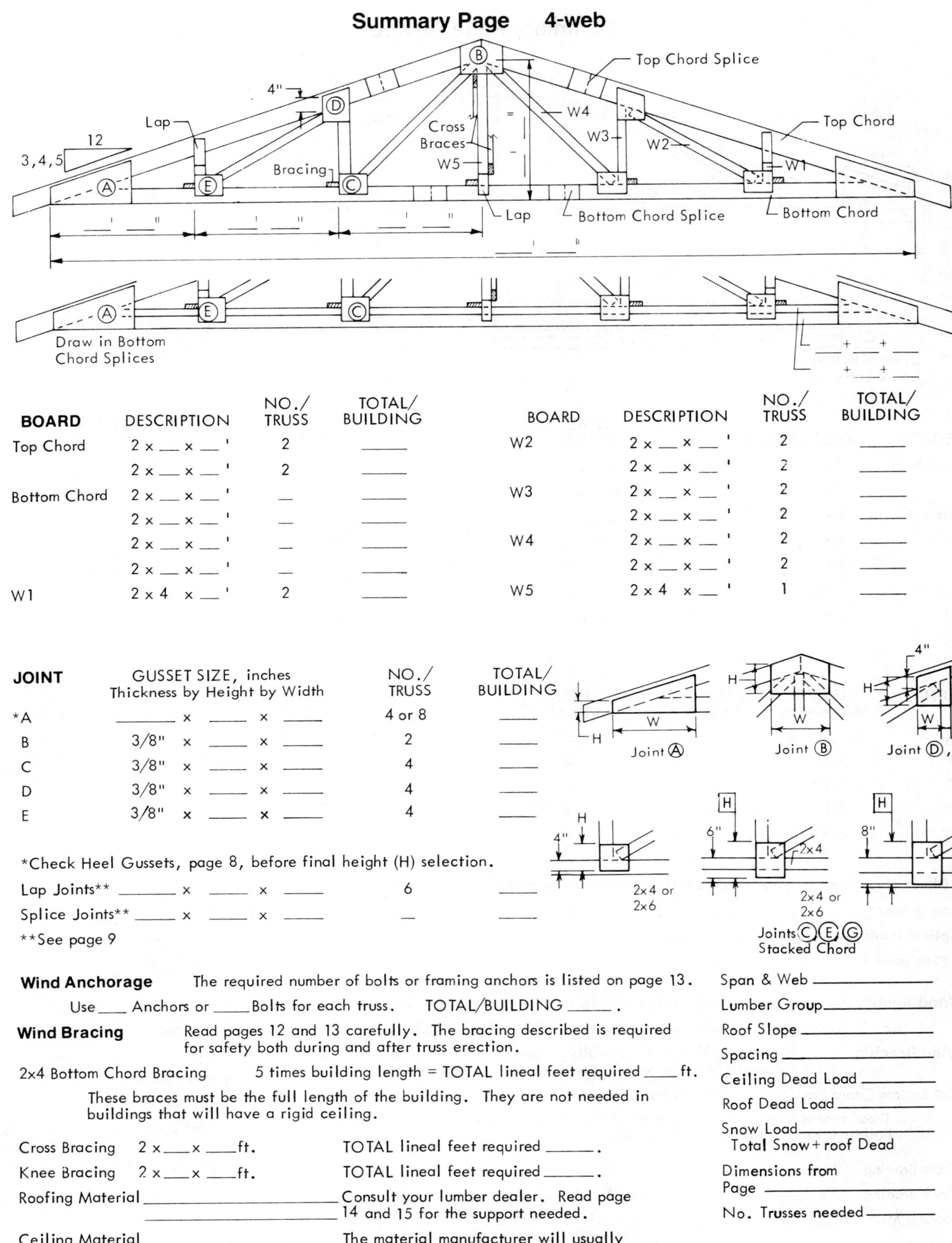

BOARD	DESCRIPTION	NO./ TRUSS	TOTAL/ BUILDING
Top Chord	2 x ___ x ___ '	2	______
	2 x ___ x ___ '	2	______
Bottom Chord	2 x ___ x ___ '	—	______
	2 x ___ x ___ '	—	______
	2 x ___ x ___ '	—	______
	2 x ___ x ___ '	—	______
W1	2 x 4 x ___ '	2	______

BOARD	DESCRIPTION	NO./ TRUSS	TOTAL/ BUILDING
W2	2 x ___ x ___ '	2	______
	2 x ___ x ___ '	2	______
W3	2 x ___ x ___ '	2	______
	2 x ___ x ___ '	2	______
W4	2 x ___ x ___ '	2	______
	2 x ___ x ___ '	2	______
W5	2 x 4 x ___ '	1	______

JOINT	GUSSET SIZE, inches Thickness by Height by Width	NO./ TRUSS	TOTAL/ BUILDING
*A	______ x ______ x ______	4 or 8	______
B	3/8" x ______ x ______	2	______
C	3/8" x ______ x ______	4	______
D	3/8" x ______ x ______	4	______
E	3/8" x ______ x ______	4	______

*Check Heel Gussets, page 8, before final height (H) selection.

Lap Joints**	______ x ______ x ______	6	______
Splice Joints**	______ x ______ x ______	—	______

**See page 9

Wind Anchorage The required number of bolts or framing anchors is listed on page 13.

Use ___ Anchors or ___ Bolts for each truss. TOTAL/BUILDING ___.

Wind Bracing Read pages 12 and 13 carefully. The bracing described is required for safety both during and after truss erection.

2x4 Bottom Chord Bracing 5 times building length = TOTAL lineal feet required ___ ft.

These braces must be the full length of the building. They are not needed in buildings that will have a rigid ceiling.

Cross Bracing 2 x ___ x ___ ft. TOTAL lineal feet required ______.

Knee Bracing 2 x ___ x ___ ft. TOTAL lineal feet required ______.

Roofing Material ______________________ Consult your lumber dealer. Read page 14 and 15 for the support needed.

Ceiling Material ______________________ The material manufacturer will usually specify the framing to use.

Span & Web ______________

Lumber Group ______________

Roof Slope ______________

Spacing ______________

Ceiling Dead Load ______________

Roof Dead Load ______________

Snow Load ______________

Total Snow + roof Dead

Dimensions from Page ______________

No. Trusses needed ______________

Summary Page 6-web

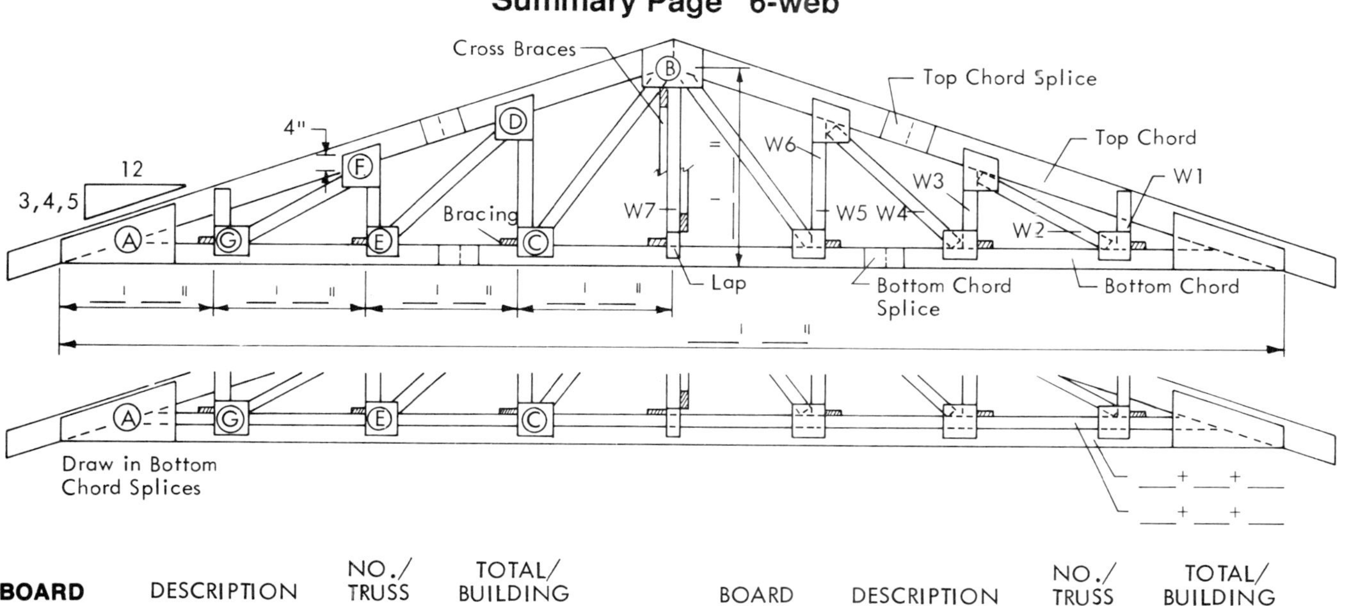

BOARD	DESCRIPTION	NO./ TRUSS	TOTAL/ BUILDING
Top Chord	2 x __ x __ '	2	______
	2 x __ x __ '	2	______
Bottom Chord	2 x __ x __ '	—	______
	2 x __ x __ '	—	______
	2 x __ x __ '	—	______
	2 x __ x __ '	—	______
W1	2 x 4 x __ '	2	______
W2	2 x __ x __ '	2	______

BOARD	DESCRIPTION	NO./ TRUSS	TOTAL/ BUILDING
W3	2 x __ x __ '	2	______
W4	2 x __ x __ '	2	______
	2 x __ x __ '	2	______
W5	2 x __ x __ '	2	______
	2 x __ x __ '	2	______
W6	2 x __ x __ '	2	______
	2 x __ x __ '	2	______
W7	2 x 4 x __ '	1	______

JOINT	GUSSET SIZE, inches Thickness by Height by Width	NO./ TRUSS	TOTAL/ BUILDING
*A	______ x _____ x _____	4 or 8	_____
B	3/8" x _____ x _____	2	_____
C	3/8" x _____ x _____	4	_____
D	3/8" x _____ x _____	4	_____
E	3/8" x _____ x _____	4	_____
F	3/8" x _____ x _____	4	_____
G	3/8" x _____ _____	4	_____

Joint Ⓐ — Joint Ⓑ — Joint Ⓓ, Ⓕ (H, W, 4")

Joints Ⓒ, Ⓔ, Ⓖ Single Chord (H, 1", 2x4 or 2x6) — Joints Ⓒ Ⓔ Ⓖ Stacked Chord (H, 1", 2x4 or 2x6, 2x4)

*Check Heel Gussets, page 8, before final height (H) selection.

Lap Joints**	______ x _____ x _____	6	_____
Splice Joints**	_____ x _____ x _____	—	_____

**See page 9

Wind Anchorage The required number of bolts or framing anchors is listed on page 13.

Use____ Anchors or _____ Bolts for each truss. TOTAL/BUILDING ______.

Wind Bracing Read pages 12 and 13 carefully. The bracing described is required for safety both during and after truss erection.

2x4 Bottom Chord Bracing 7 times building length = TOTAL lineal feet required____ft.

These braces must be the full length of the building. They are not needed in buildings that will have a rigid ceiling.

Cross Bracing 2 x ___ x ___ ft. TOTAL lineal feet required_______.

Knee Bracing 2 x ___ x ___ ft. TOTAL lineal feet required_______.

Roofing Material ____________________ Consult your lumber dealer. Read page 14 and 15 for the support needed.

Ceiling Material ____________________ The material manufacturer will usually specify the framing to use.

Span & Web ____________

Lumber Group ____________

Roof Slope ____________

Spacing ____________

Ceiling Dead Load ____________

Roof Dead Load ____________

Snow Load ____________
Total Snow + roof Dead

Dimensions from Page ____________

No. Trusses needed ____________

Summary Page 6-web

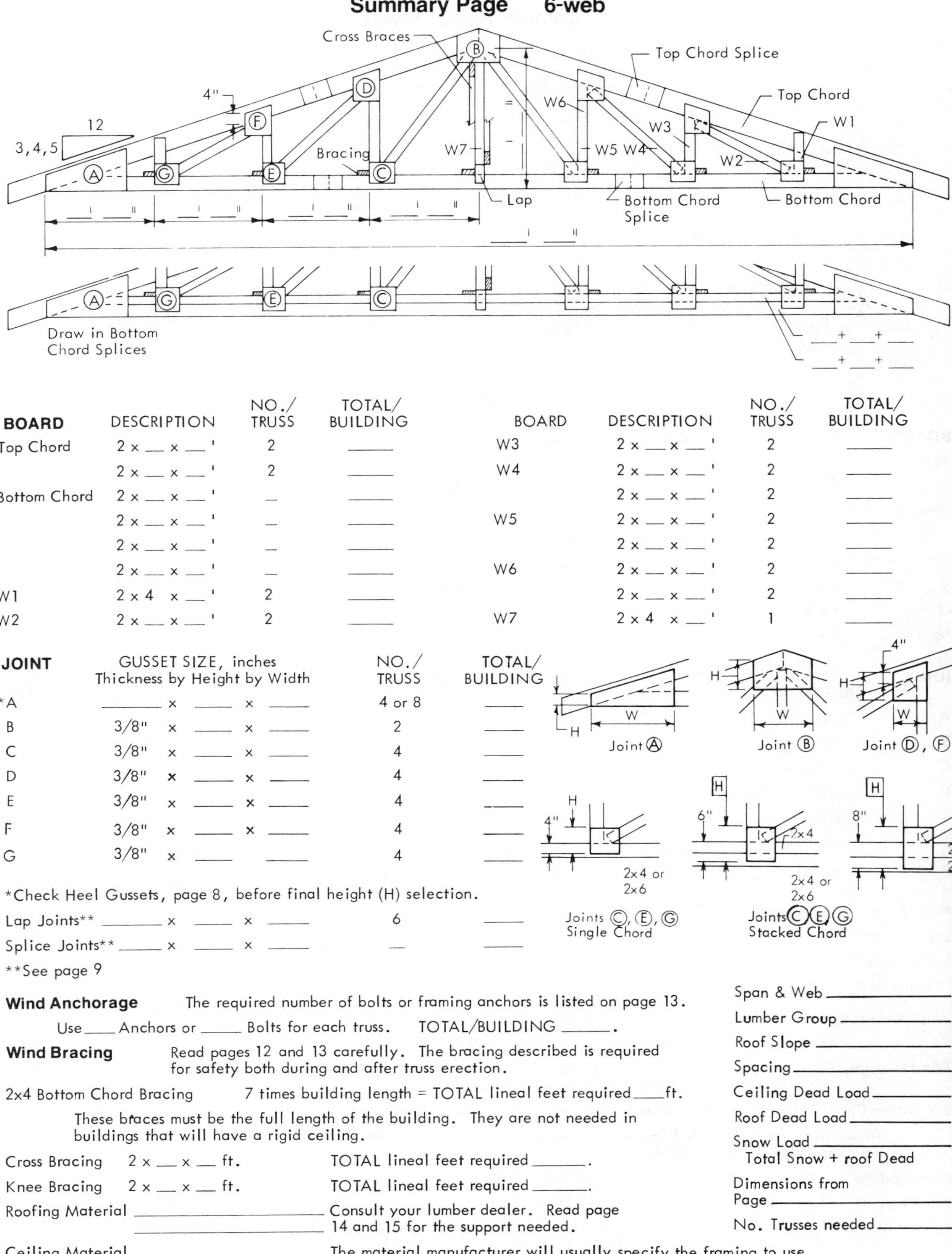

BOARD	DESCRIPTION	NO./ TRUSS	TOTAL/ BUILDING
Top Chord	2 x __ x __ '	2	_____
	2 x __ x __ '	2	_____
Bottom Chord	2 x __ x __ '	_	_____
	2 x __ x __ '	_	_____
	2 x __ x __ '	_	_____
	2 x __ x __ '	_	_____
W1	2 x 4 x __ '	2	_____
W2	2 x __ x __ '	2	_____
W3	2 x __ x __ '	2	_____
W4	2 x __ x __ '	2	_____
	2 x __ x __ '	2	_____
W5	2 x __ x __ '	2	_____
	2 x __ x __ '	2	_____
W6	2 x __ x __ '	2	_____
	2 x __ x __ '	2	_____
W7	2 x 4 x __ '	1	_____

JOINT	GUSSET SIZE, inches Thickness by Height by Width	NO./ TRUSS	TOTAL/ BUILDING
*A	____ x ____ x ____	4 or 8	____
B	3/8" x ____ x ____	2	____
C	3/8" x ____ x ____	4	____
D	3/8" x ____ x ____	4	____
E	3/8" x ____ x ____	4	____
F	3/8" x ____ x ____	4	____
G	3/8" x ____ ____	4	____

*Check Heel Gussets, page 8, before final height (H) selection.

Lap Joints**	____ x ____ x ____	6	____
Splice Joints**	____ x ____ x ____	—	____

**See page 9

Wind Anchorage The required number of bolts or framing anchors is listed on page 13.

Use ____ Anchors or ____ Bolts for each truss. TOTAL/BUILDING ____.

Wind Bracing Read pages 12 and 13 carefully. The bracing described is required for safety both during and after truss erection.

2x4 Bottom Chord Bracing 7 times building length = TOTAL lineal feet required ____ ft.

These braces must be the full length of the building. They are not needed in buildings that will have a rigid ceiling.

Cross Bracing 2 x __ x __ ft. TOTAL lineal feet required _____.

Knee Bracing 2 x __ x __ ft. TOTAL lineal feet required _____.

Roofing Material __________ Consult your lumber dealer. Read page 14 and 15 for the support needed.

Ceiling Material __________ The material manufacturer will usually specify the framing to use.

Span & Web ________

Lumber Group ________

Roof Slope ________

Spacing ________

Ceiling Dead Load ________

Roof Dead Load ________

Snow Load ________
Total Snow + roof Dead

Dimensions from Page ________

No. Trusses needed ________

NOTES

NOTES

NOTES

NOTES

NOTES

NOTES